The Wl
of God

1801-2027

How solar eclipse paths have defined the geopolitical
conflicts of the modern age and what they foretell for
Mankind's future.

Keith Magnay

Tzaddi Books
London

The Wheals of God
Produced and published by
Tzaddi Books
London

First published in 2013
Copyright Keith Magnay, May, 2013

ISBN No. 978-0-9576302-0-8

thewhealsofgod.com

All Solar Maps graphics generated by Solar Fire v8 are reproduced here with the kind permission of Astrolabe Inc. (www.alabe.com)

Front cover picture: Antoine Caron's haunting "Astronomers Studying an Eclipse" of 1575. The French Court painter would have seen the eclipse which occurred four years earlier a few months before the pivotal Battle of Lepanto when European and Ottoman fleets fought over mastery of the Mediterranean Sea.

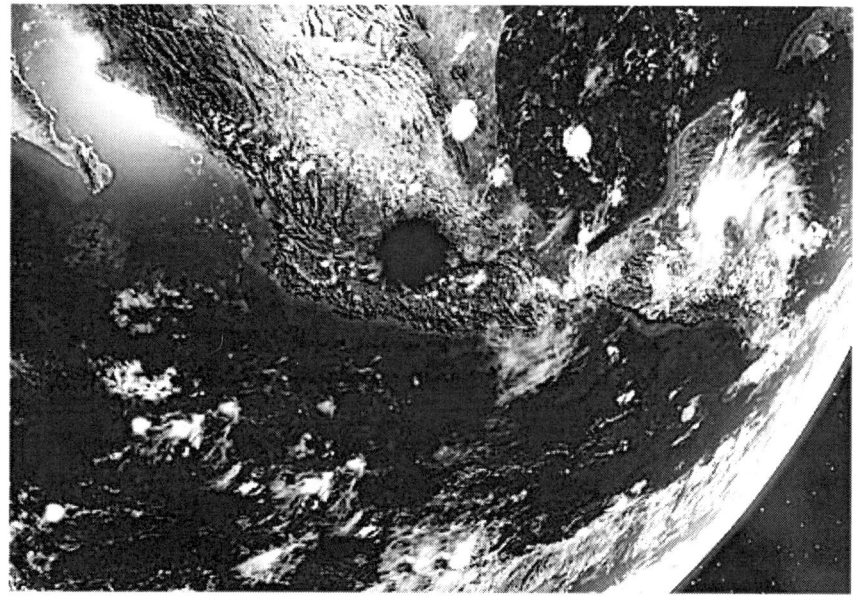

The ominous shadow of an eclipse falls over Mexico as shown from a NASA satellite.

The strange rays coming from an eclipse. Photo:NASA

What they might have said if they had reviewed this book.

It's history, Jim. But not as we know it."
Mr Spock, Vulcan First Officer, Starship Enterprise

"The truth is in here."
Fox Mulder, Special FBI agent, the X-Files

"Reads like one of my screenplays."
Quentin Tarantino, Film Director.

"It's like Fifty Shades of Grey, only on an historical scale."
E. L. James, contemporary author and connoisseur of sado-masochism.

"The period of breakdown of civilisations is characterised by eclipses."
Arnold Toynbee, author of A Study of History.

"Well it was an eclipse that proved my theory of relativity."
Albert Einstein, Astrophysics Theorist

"The equivalent of Heston Blumenthal in food for thought."
Tom Parker-Bowles, food writer.

"This book will be feared rather than loved."
Niccolo Machiavelli, Renaissance Political Consultant

"Magnay's prose is like a fire-fight."
Andy McNab, SAS author

"Beats my calendars."
Hunab Ku, Mayan Day-Keeper

"It seems to be saying that states will never have rest from evils until politicians become astronomers and astronomers become politicians."
Plato, Athenian Philosopher.

"There is more between heaven and earth than is dreamt of by most historians."
William Shakespeare, Elizabethan Playwright.

"Don't blame it on sunshine.
  Don't blame it on moonlight
  Don't blame it on bad times
  Blame it on the eclipse.
Michael Jackson, deceased singer.

"When sun and moon mate in sky.
  Fighting dragons to earth do fly."
Confucius, Chinese Sage.

"As if poverty, old age, disease and death were not enough."
Gautama Buddha, Religious Ascetic.

"Not a lot of people know what's in this book."
Michael Caine, actor and autobiographer.

"This book is poisonous."
Vladimir Putin, President of Russia (again).

"I still have faith in God."
Job, Afflicted Monotheistic Believer

"Poppycock, balderdash and pernicious piffle"
Prof. Richard Overy, Editor, Times Complete History of the World

"In the year of two popes
Eclipse revelations will overshadow all hopes
A discovery of the source of political ire,
Amidst foretelling exceeding dire"
Michel Nostradamus, 16th Century seer

"The scariest, Scary-Book in Scaryland."
Ben Elton, Comic Writer

"Every so often a book like Charles Darwin's Origin of Species, Sir Isaac
Newton's Principia Mathematica and my own book, The Selfish Gene, comes
along and transforms our understanding of the natural world and ourselves. Well.
This isn't one of them."
Richard Dawkins, pro-scientific atheist

About the Author

This work is the result of a peculiar combination of the author's interests which have informed his adult life. It began with an "Awe of God" experience whilst studying history and sociology at the University of York, England, some 40 years ago.

An epiphany about "being", coupled with scientific training, led to a life-long research of religions and philosophies from a perspective unbiased by culture and facilitated by a career in domestic and international journalism.

Along the way he investigated astrology, kabbalah, hermeticism, shamanism and Rosicrucianism, much of which entails the deployment and transcendent interpretation of symbols.

The Wheals of God is the product of his historical and religious studies, an understanding of the hermetic dictum "As Above, So Below" and his experience in interpreting symbols.

It was this background which enabled him to unveil the true mystery of history and make the connections between the lines of eclipses and mankind's past and future travails.

This book gives thanks to:

the professors who guided my path of historical and cultural inquiry
G. W., G. L. and R. R.;

the Gemini muses who guided me through love to wisdom
F. S., E. M. and C. R.;

the initiators who guided me from darkness to light
P. C., J. M. and C. C.;

the shamen who guided me to the realms of the hidden knowledge
P. A., R. and A. C.

and to Rick Falconer for sprinkling his stardust.

"Poor men. What terror is this that overwhelms you so?
Night shrouds your heads, your faces down to your knees.
Cries of mourning are bursting into fire, cheeks rivering tears.
The walls and the handsome crossbeams dripping dank with blood.
Ghosts, look, thronging the entrance, thronging the court,
go trooping down to the realm of death and darkness!
The sun is blotted out of the sky. Look there.
A lethal mist hovers above the earth!"

Homer's Odyssey.

"Therefore out of Light and Water has God created all Things."

Golden Chain of Homer

"The Axis in the Universe is like a King on his Throne
The Cycle in the Year is like a King in the Province
The Heart in the Soul is like a King in War."

Sepher Yetzirah

"And even the very hairs of your head are all numbered."

Matthew 10:30

"For what we do presage is not in grosse
For we be brethren of the Rosie Crosse
We have the Mason word and second sight
Things for to come we can foretell aright."

Henry Adamson, Dean of the Merchant Guildry and Provost of Perth, 1638

From Darkness comes light.

The Gift to Mankind from Love Destroyed

She started sweetly and pragmatically enough.

"It was too cloudy to see the eclipse," her email opened, giving no indication of what was to follow.

I was pleased to hear from her. A goddess was communicating directly to me from her home in Japan, 6,000 miles away.

Her writings usually emitted a sense of deliberation and purity of sacredness as if she had banished the mundane world when she sat at her desk, meditating in the quiet of the night.

But this time, her first communication to me after the eclipse, it was different.

As her message progressed she led me into an abyss, gradually revealing an implacable anger towards me, finding fault, declaring that unintended circumstances and alleged lacks of integrity were deep karmas between us, deliberate acts of my betrayal and sabotage.

She knew she had engendered deep feelings in me, enabling me to reach, once again, to the burning source of love in the universe, the wellspring of Creation. But now she was transformed from beauty, grace and exquisite manners into a Fury issuing a mocking, pitiless rejection. I could do no right. She was condemning me, absolutely, to swing from Odin's Tree.

The woman I had enshrined in a cocoon of Platonic love was implacable, cruel, a pure crystalline severity.

She had become the truth of my observation that those who advocate love alone for the world invariably come to a martyr's end, Jesus Christ, Mahatma Gandhi, Martin Luther King and John Lennon, amongst them.

I had known that my inner life would be racked when she left England three years earlier, untimely ripped from her work in temple and that I would miss her awfully.

As I had anticipated her forthcoming moving on in life I could only brace myself to surf the intense, wasteful emotions that would follow, producing a seepage of both tears and enjoyment of life.

I felt locked into a fateful experience, one I had premonitioned when her application came in and I had read her horoscope to test her suitability for a course in alchemy and ascension. I knew her astral make-up could stir me before I had even met her.

Sustaining communication with her was always going to become more difficult over time and it could be maintained only if she felt similarly, which I doubted.

But I had been there before and knew that she was the third and concluding Beatrice of my life.

Sophia, the goddess of wisdom, was working through her and I anticipated my soul's fiery consumption would produce gifts of wisdom, knowledge and understanding just as I had been led by two earlier muses from a focus on sport to one on literature and from a perplexed existentialism to the certainties of astrology.

In our correspondence she took me into her own ascended realms, the teacher learning from the pupil. She confided that she had incarnated to assist humanity on its path and gave me four dates over a period of more than a decade in the '90s and early '00s on which she had, in her dreams, worked in the astral and opened up gateways to clear the path for Mankind to discover the Hidden Knowledge.

I was astonished to see how the planetary configurations of these dates reflected exactly the process of first an initiation, followed by a structuring, a catalytic speeding up and finally an earthing.

I had warned her of the eclipse that would pass directly over her home on May 20, 2012, that it would herald a disaster for Mankind and especially for Japan. She re-acted with a resigned indifference.

I knew that I was flailing around, unable to impress her with my understanding of the eclipse's true import. I had to provide her with more evidence, to demonstrate exactly how eclipses have affected us over historic time.

As the high priestess of sublimation, she was goading me to search more deeply and, circuitously, to assist her in the process of bringing the hidden knowledge to Mankind.

Through her harsh treatment of me she was bestowing the Gifts my feelings had promised as I was provoked to carry out the research that has produced the astonishing and powerful revelations in this book.

Like Prometheus I had been enabled to steal from the Gods a huge benefit of understanding for humankind. Unlike Prometheus it was not my liver that would be eaten every day but my heart.

CONTENTS

CONTENTS

CONTENTS

CONTENTS

xvi

CONTENTS

Introduction

At the centre of the historical process lies military organisation and the willingness of the leaders of political bodies to shed both the blood of their own people and that of other peoples.

Weakness in numbers, resources and technology has been punished time and time again throughout history as has overstretch, lack of diplomacy and manoeuvre and the underestimation of opponents.

Natural history is the response of creatures to meeting the requirements of food, reproduction and territorial base. So too it is with us human beings, although in our case, the fault lines in the struggles between ourselves are marked by an additional natural phenomena, that of solar eclipses.

It is the thesis of this book that solar eclipses are the drivers of the historical process, the timers of God's chess moves against mankind, the way he expresses his will. As such they are the signposts of tangible shifts in the power and control of segments of the earth's resources.

Significant human conflicts either arise or are exacerbated in their wake, literally testing the power of those who fall under their shadow. In the process of revealing where political strength really lies they also parallel terrible sufferings.

Eclipses are like whippings that leave historical reddenings and bruisings on Mankind's skin. These wars, rebellions and changes of territorial and constitutional status may be viewed as the Wheals of God.

This book's story line is a tough tale, both dark and pessimistic, for which I apologise and ask you to understand that I am only reporting what I have found. It is also one from which Mankind can receive light and optimism. After all at night-time we can see the stars and the immensity of the universe and more clearly experience the overwhelming power of Creation. It is my sincere wish that, through the understandings found here, Mankind can learn its interconnectedness with the cosmos and solar system and politicians, religious leaders and global government agencies can both make this connection and anticipate the future locations of their beneficial attention and energies.

Although it is a tall order scientists now have an opportunity to study the physical and electromagnetic effects of eclipses in a bid to counteract them and potentially end war on earth.

In the meantime this book is also an appeal for understanding and wisdom about the human condition from a macrocosmic perspective, one which urgently requests a more spiritual approach to how we conduct our lives, share the earth's resources and attain harmony between ourselves, knowing that we will regularly

1

experience severe astronomical provocations.

This view has particular urgency today because of the development and spread of weapons of mass destruction, including nuclear, biological, chemical and electromagnetic. Our economic, technological and judicial civilisation is very fragile and we have a very long way to fall should serious conflict break out at a global or even a regional level.

This is an history book, but it was my study of mundane astrology that first sparked my interest in the subject. The leading mundane astrologer Charles Harvey, drawing upon the work on eclipse paths of Charles Emerson, Al Morrison and others, suggested that eclipses signified the rise and falls of empires and also connected pivotal individuals in history.

He observed that eclipses close to their births marked the lands over which Alexander the Great, Mohammad and Karl Marx would conquer or greatly influence.

For example the one around the time of the birth of Mohammad was believed to have traced out the line of the countries which would become Islamic i.e. North Africa, Arabia, Persia and North India.

This eclipse seems to have both defined and confined Islam in an East-West band around 10 - 40 degrees North. Islam has struggled to become dominant elsewhere.

It made me think it must have been very powerful to have paralleled such a major cultural shift in Mankind's history. After all Islam is passionately adhered to by a quarter of the world's population today, nearly 1400 years later.

The history of Islam shows that it grew in many instances by conversion through the sword and that much blood was spilt as it achieved its supremacy. It was as if the eclipse had demanded that man fight each other, like Cain and Abel. Was this a characteristic of eclipses?

Modern computing has enabled the calculation of where eclipses fell and will fall and the capability to show them graphically. Solar Fire's Solar Maps programme, produced by Esoteric Technologies for Astrolabe Inc, has 400 years of eclipse lines from 1801 to 2200. This provided me with the ability to investigate if other eclipse lines had any bearing upon history albeit limited to the modern period.

First I looked to see if there was an eclipse around the First World War. Was there an eclipse at its inception close to where the fighting took place?

Well yes there was. Just two weeks after the declaration of hostilities, on August 21, 1914.

This eclipse's path was as close a demarcation of the Eastern Front as you could get and especially of the peel-backed boundary of Russia following the 1918 Treaty of Brest-Litovsk.

Fighting was to move up to and criss-cross this eclipse line for the next thirty years devastating the cities under its path.

The Treaty of Brest-Litovsk confirmed the birth of the world's first communist state in Russia and its withdrawal from the war with Germany. Communism was to become a dominating world historical force for the ensuing 72 years.

I then found that the Western Front too had been indicated by the earlier eclipse of April 17, 1912. This cut through Belgium and France in a line which anticipated the pivot of the attack of the Kaiser's troops in their stalled drive towards Paris two years later.

I then turned my attention to the Second World War. Initially I was unable to find any significant eclipse line around a 1939 start until I realized that I was being UK-centric.

Objectively, it dawned upon me, it had really begun much earlier during the period of appeasement which had been adopted as Hitler remilitarised the Rhineland and took over the Saarland, Austria and the Sudetenland ahead of the attack on Poland in 1939.

Sure enough the eclipse of June 19, 1936, set the scene for the Second World War covering all its multi-theatres of Europe, Russia, China, Japan and the Pacific as well as highlighting the Mediterranean and North Africa. It occurred shortly after Germany retook the Saarland and a year before Japan's full scale invasion of China.

For nearly half the twentieth century the World was in the grip of a Cold War between the liberal democracies led by the USA and the countries of the Soviet model sponsored by Russia.

Was there an eclipse which reflected this global contest? Well, yes there was.

What about the Korean War? Yes, that too and it explained why it was a war between two halves of that peninsular country. And Vietnam as well, explaining why it was divided and also why it was so protracted.

I also found another eclipse which outlined the present conflict between the West and militant Islam.

When I presented my initial findings to friends and near associates, one or two of the sceptical ones responded by saying that there were so many eclipses that there was always going to be an eclipse around an event.

Not entirely true, of course, as the probability of the connections I found being entirely random even at that stage was highly unlikely.

However this prompted me to adopt a second approach and conduct a comprehensive survey of the timings and courses of all the eclipses over the past 213 years and further on.

The partial results of this are to be found at the back of this book. I found that quite a large number of eclipses are over places where humanity is not situated such as our oceans, deserts, forests and the frigid zones and therefore had no historical impact.

Others were of a regional nature over recognisable smaller scale political entities but were not immediately identifiable by me as being connected to specific historical events in those countries.

However the historical research into some of the less obvious eclipses proved very fertile and I found even more sets of connections which fitted many of the key events identified by historians of the modern period.

In this category are the sections on the scramble for and decolonisation of Africa, the Wars of Independence in South America as well as the specific sections on the British Empire, South Africa and the Sudan.

This approach also threw up, completely unexpectedly, the Asian Independence eclipse of 1944 and the rather strange 1887 seeding of the dark rulerships of the twentieth century war-mongers, Germany, Russia and Japan.

Of course this book will incite many different re-actions depending on the consciousness of the individual reader just as some seeds fall on stony ground and others on fertile.

I do not claim to be 100% correct in all the connections I have made and I am sure that other connections can also be added in due course, but I do believe that the core thesis of the connection between eclipses and Mankind's conflicts and atrocities is one on which light has now been shed and that this understanding should be taken on board by those who are in significant positions of global leadership.

I do not profess to know exactly how eclipses bring about the correlations I have unveiled but I would like to propose that it could be related to the properties of the light that occur during an eclipse.

Homer wrote that a "lethal mist hovers over the earth" around an eclipse. Contemporary photographs from outer space both of an eclipse's shadow on the earth and of the moon occluding the sun induce a similar response.

From outer space the shadow of an eclipse appears like a dark abyss, a bottomless, black hole. It is as if a bullet has been shot through the earth. According to one commentator on a NASA photograph of a solar eclipse, looking towards the sun the solar corona appears to be shining with "a strange, silvery light around the black night side of the moon."

Introduction

He goes on:"When the last rays from the sun appear behind the moon the beautiful diamond ring effect can be enjoyed for a few seconds. In a moment the solar corona, the outer solar atmosphere, emerges. Towards the horizon in all directions there is a strange reddish light. The temperature falls, insects and birds stop making sounds and the landscape changes."

These observed "strange" fluctuations in light during an eclipse whether silvery, reddish or dark may be a fruitful area of research for scientists wishing to understand the relationship between the human brain and electro-magnetism

Some of US President Obama's recently announced funding of research into how the brain works may be usefully deployed in this area. It may be that such a study will help us understand why conflict and atrocity have a tendency to follow after an eclipse.

According to the Golden Chain of Homer, light is fundamental to the composition of the matter of the universe in which we have our existence and it does not take too much stretch of the imagination to believe that its distortion is likely to have a primal effect on human beings.

The physical impact of solar eclipses on the earth's air movements have already been monitored and have revealed that waves similar to those generated in water by ships are stimulated. These bow and stern waves churning in the upper atmosphere and ionosphere are the result of the differences in temperature between sections under shadow and those remaining in sunlight and have been described as having the force of being cut through by a ship 560 miles long travelling at 200 miles per hour. That's a very large ship and it is not surprising that it has a tremendous impact!

As a child I grew up amongst farm workers, some of whom had lost brothers in the Great (First) War. From them I heard the acquiescent lament of many of the Tommies faced with having to make a nonsensical sacrifice of their lives:
"Ours is not to reason why.
Ours is but to do and die."

It is my hope that this book, although too late for them, finally, provides the astronomical reason why and that the astrophysics of history's evils have been brought to light.

The Astronomy of Solar Eclipses

Anyone who has stood underneath the path of a total solar eclipse will know what an awesome experience it is. You can feel the power and the majesty of the solar system as its components encircle each other.

The alignment of the sun, moon and the earth takes place at a distance of over 93 million miles and involve three spherical bodies of different sizes and orbits, suspended in space and moving, not just in ellipses but also vertically in sine wave motions as well.

It is not surprising to appreciate that these occasional bull's eye shadows fall on the earth in seemingly irregular locations.

The timings of solar eclipses are restricted not only to new moon conjunctions but also only to when the moon passes the earth's ecliptic, which it does twice a year.

These points in the annual cycle are known as the moon's north and south nodes or dragon's head and dragon's tail. They are similar to the earth's spring and autumn equinoxes when the sun passes the earth's ecliptic (equator) but are not fixed, moving retrograde against the zodiac in an 18-year plus cycle.

Co-incidently the sun is around 400 times larger than the moon but is also 400 times further away from the earth. This means that they appear to be of equal size and the moon can completely blot out the sun as viewed from the earth.

The shadow cast by the moon is known as its umbra. It is total in a narrow line but is partial in a broad swathe either side. It moves across the earth in an easterly direction at the net speed of the moon compared to the rotation of the earth.

It can be seen for about three minutes along any point of its totality, but, very rarely, can last for up to seven minutes. At the outset the moon takes about an hour to blot out the sun's light which is reduced to what has been described as a strong moonlight.

Eclipses fall into four types depending on their visibility from the earth. The first is a total eclipse when the sun is completely obscured by the moon. The partial parts of a total solar eclipse are known as the moon's penumbra.

An annular eclipse occurs when the moon is closer to the earth and does not cover the sun completely, creating a doughnut effect with the central darkness encircled by a solar ring.

A partial eclipse is when there is no line of totality falling on the earth and the sun appears like a crescent. The amount of occlusion can vary considerably, from just over 0% to just under 100%.

The fourth type, hybrid, varies between totality and annular during its course.

According to NASA, during the 5,000 years between 1999 BCE and 3,000 AD some 11,898 eclipses will have occurred, an average of marginally under 2.38 every year. Of these 4,200 (35.3%) will have been partial, 3,956 (33.2%) annular, 3,173 (26.7%) total and 569 (4.8%) hybrid.

The line of totality is roughly visible in a band 150 kilometres or 94 miles wide. This is 1/130th of the distance between the North and the South Poles. By definition solar eclipses can only occur in the half of the earth that is in daylight. Travelling in a west-to east direction they usually extend along 140 degrees of longitude or 7/18ths of the earth's circumference.

This means that the chances of the line of totality passing overhead at any point on the earth is roughly 1 in 500 or once in every 208 years.

In terms of human habitation of the earth only 1 per cent is covered by our species' settlements. Nearly 71% of the earth is covered by water; 11% is wilderness and 9% is desert or near desert. Antarctica is 3% with the remaining 6% being agricultural.

It should not be surprising therefore to be informed that many eclipses fall over remote, uninhabited locations and do not influence the historical process. Nowadays, with globalisation, when travelling over land they are likely to pass over a political entity and the larger the country the more likely it is to be overshadowed and therefore to experience conflict.

Clearly countries that have a long north-south border are most likely to be affected. Amongst them are the USA, China, Russia, Mexico, Chile, Argentina, Brazil, Japan and Vietnam; Europe too, taken as a whole. Larger countries in Africa such as South Africa, the Sudan and the former Zaire are also more vulnerable.

This study is especially concerned with the umbra or line of totality of eclipses but some partial eclipses are included when the correspondence between them and significant events is particularly strong.

How to Read Eclipse Path Lines

The eclipse path graphics are shown with their date and a heading connecting them to an event close in time and space. In each case the length of time between the eclipse and the power-changing event is also given in the body text.

The associated events vary in depth of horror and in the time they take to have fullest effect. The general pattern though is within two to three years after the eclipse but their impact can be for decades often being compounded with later eclipses.

Some eclipses are very exact in terms of the location they affect such as those around the Korean, Vietnamese, Sudanese and Georgian wars whilst others appear to pick out an increase in severity of an already existing conflict, the ones around the Boer War and the Indian Mutiny being examples of this.

The events cited are usually a war, a rebellion or a change or challenge to a constitutional or territorial status. They invariably involve significant and cruel, injudicial loss of life. Some are associated with ideas and others with the birth or the death of an historic figure.

To examine an individual eclipse path make a list of the countries that the line of totality passed over, drilling down to individual towns to find out if it went over a capital city or other significant centre. Countries alongside, which experienced over 50% totality, should also be noted as a secondary focus.

Then take a look at the history of those countries for up to three years later to identify any precipitous events or changes in tone. Reflect on the seriousness of this event for the individual country, its region and the world.

Then take a look at the relationships between the connected countries in the immediate future and then longer term. Did they share a common theme? Were they friends or allies? Was there an imbalance in the power relationship between them exploited by one or other of them?

After this analysis ask yourself if it is an "event" eclipse or an "era" eclipse? Can it be judged to have seeded a long term process and therefore connect to other eclipses or was it a one-off?

These are the approaches that I have used with some genuinely surprising results. As an add-on in your contemplation of this work and, in part, a justification of this line of research, I will reference the methodology of the subtle sages of Jewish Kabbalah.

They use four levels of interpretation of biblical texts known as Pardes, written as PRDS in Hebrew and paradise in English.

Each letter stands progressively for a way of understanding a writing as you can from the lines of the eclipses in this book.

P stands for Pshat which means simple or literal interpretation.

R stands for Remez, which means "hint" or allegorical interpretation.

D stands for D'Rash, which means to investigate, seek out or expound as in Aggadic or Talmudic interpretations.

S stands for Sod or secret (kabbalistic) interpretation.

In this instance Pschat is the physical fact of solar eclipses, the actual time they took place and the locations they went over as shown in the eclipse path maps. Those of a literal mindset will take them at their face value, viewing them as nothing more than an occasional phenomenon with no more import than the fact of their occurrence.

Remez, in this instance, is their connection to co-incidental historical events derived from well documented timelines and historical works. This requires the capacity to find correspondences between areas that do not seem, at face value, to be connected.

D'Rash is their interpretation as part of the long term historical process. The building up of related historical events compatible with the main thesis of the malevolence of eclipses is a form of expounding on the first two levels of this system of multi-layered understanding. Hence the grouping of eclipses into chapters around the world's regions, ideologies and superpowers.

Sod is their import as agents of Divine will. As you will see what is outlined here can be used for historical divinatory purposes so that future areas of conflict can be identified ahead of time. God as the engenderer of history is being revealed as confined by cycles of time and astronomical measurements.

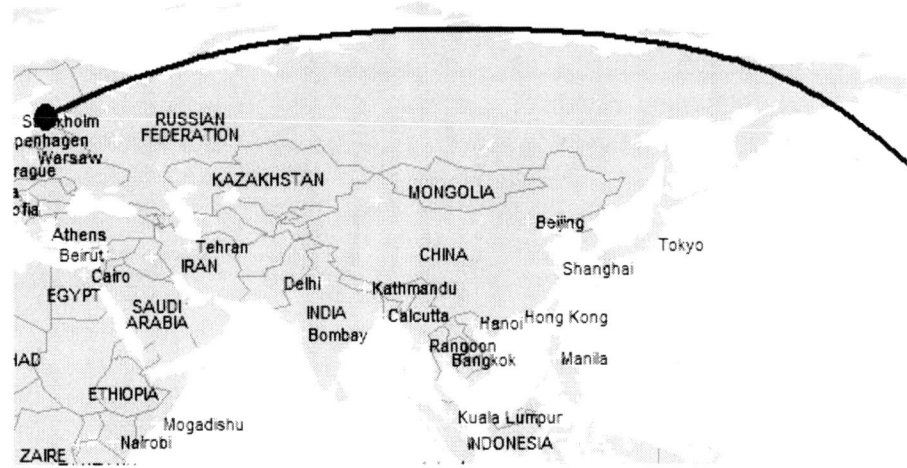

This eclipse shows the path of totality or umbra for July 22, 1990 just ahead of the collapse of the Soviet Union that it encircled. The lines of partiality around the same eclipse, covering the whole of the affected region, are shown below.

The majority of the graphics used just show the line of totality but it should be borne in mind that partial views of eclipses are visible from a far greater area, in this case the whole of the then Soviet Union which was about to disintegrate. Please note that the 5 at the end of the eclipse line and all others in this book should be taken as 50.

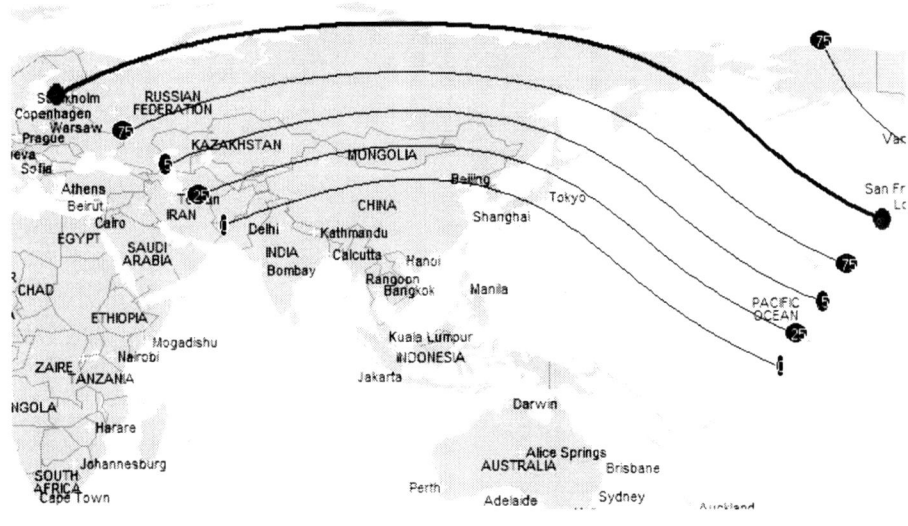

Twentieth Century Wars

How solar eclipse paths marked the locations, antagonists and sometimes even the victors of the global wars of the 1900s and early 2000s.

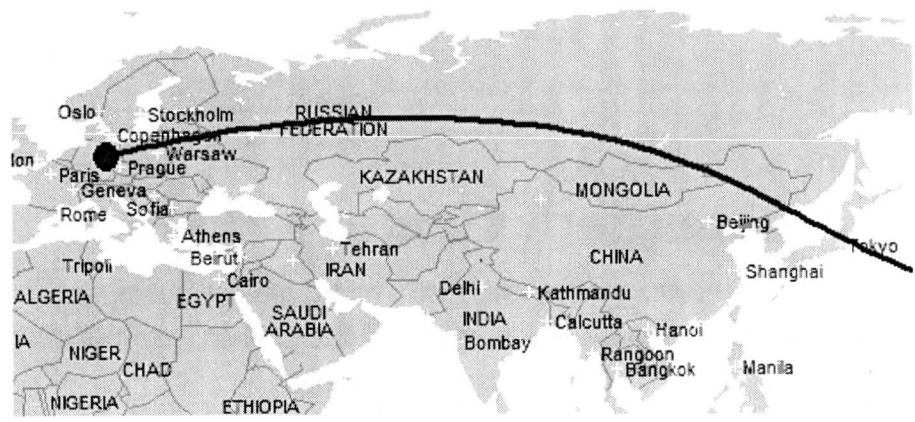

The Three Aggressors Solar Eclipse of August 19, 1887

This might seem to be a strange date to start the total wars of the twentieth century and it surprised me too when I came across it.

Originally I was going to bracket it in the European dynastic section until I realised its broader and longer term significance.

This eclipse emerged over Berlin and went directly over Moscow and Tokyo, the capitals of three of the major 20th Century belligerents, Germany, Russia and Japan, as if they were being marked out to become warmongers?

Around this time these three countries, seeking to emulate and muscle in on the established empires of the Western European maritime states all became predators on the Eurasian landmass and sometimes further afield. Significantly they were all willing to fight against industrialised nations rather than those with primitive weaponry.

Germany was the instigator of the First and, with Russia, the Second World War in Europe; Japan of the Second World War in the Pacific and Russia of the Cold War conflicts including the Korean and Vietnamese Wars.

From this time they were to be involved in constant alliances and tugs-of-war between themselves and also the Anglo, Franco and American axis.

All three embraced authoritarian militarist political systems either of the right or the left and used war to expand their control and defense of world resources.

Each of them expended millions of the lives of their own countrymen and took the lives of millions of those from other countries especially their neighbours through a pick 'n' mix of genocide, gangsterism and absolute subjugation.

This eclipse appears to have had a long term effect and to have sown a particularly ruthless, even criminal, mind-set in the Chancellery, Kremlin and Imperial Palace of these three countries especially in their militaries with disastrous consequences for Mankind for decades to come.

This is not to place the blame on them completely or to excuse the Anglo-American and Western European power bloc which spent much of the twentieth century trying to contain the worst excesses of these three brutalising powers on the world stage.

This bloc had its unpalatable side too as American Indians, Indian mutineers, Boer farmers, Algerians and Black American slaves can testify but dictatorship was not one of its faults.

The galvanising effects of this eclipse prompted Germany to become competitive in Africa propelling the scramble for lands in the continent by Europe's imperialists a few years later.

Over the next sixty years Germany's ambitions gave high anxiety to its fellow European states many of whom were to be overrun by it more than once.

The following year, 1888, is known as the Year of Three Emperors in German history. It was a pivotal transfer of power from the avuncular 91-year-old Kaiser Wilhelm I who depended on the shrewd statesmanship of his "Iron Chancellor", Otto von Bismark.

It passed briefly to his war-hero son Frederick William, a liberal married to a daughter of Britain's Queen Victoria. He never spoke during his short 100-day reign, dying of cancer of the larynx.

The poisoned sceptre passed to his wilful son, Wilhelm II, whose illiberal tendencies, pride and desire for glory were to bring about an abyss in the storyline of European civilisation and an end to his own dynasty. He was to launch his country on a path that took it from a position of respect and prestige to one of opprobrium and disgust.

To rework an observation of Henry Kissinger's, Kaiser Bill was correct in understanding that Germany should be regarded as the foremost power in Europe but incorrect, as was Adolf Hitler, in presuming that it could also be the first power in the world.

Most especially his sacking of Bismark led to Russia joining up with France making a war on two fronts more likely.

Japan's rapid modernisation programme since the Meiji restoration twenty years earlier was also giving it an appetite for conquest. After hiding away for over 200 years it had undergone a steep learning curve adopting the technologies and systems of the furtherest advanced European states.

On sea power it absorbed the most from the British and French navies and for its army it turned to its eclipse buddy, the Prussians. It also brought in a European constitution at this time.

Japan had kept social order through the samurai, a warrior caste who had an heartless right of life and death over civil populations. Immediate decapitation for even the slightest indication of disrespect was standard.

The samurai were pushed aside by the mechanisation of long range weapons but their ethos found its way into the Japanese Army and was to be turned on the peoples of East and South East Asia.

It was at the time of this eclipse that the Japanese Army expanded from being a civil defense force capable of beating off an invader to become an aggressive tool of imperialism. The switch of structure from garrison to division was an indication of this change of emphasis and intent to project power internationally.

So fanatical was Japan in building up its armed forces at this time in direct competition with China that it almost choked on the financial costs, sending the country's economy into a tailspin.

Russia at this time was an asylum of paranoids, a scorpion's nest of whispered plots, agent provocateurs and police informants. True, serfdom had been abolished but a liberal Tsar had been thanked for his benevolence by assassination and a counter revolutionary mentality now dominated the Russian court.

Alexander III, fearful he would go the way of his father, hid in his palace and surrounded himself with his Okrana, "guardian" secret police who were to become the model for the Cheka of the Bolsheviks and the KGB of the Cold War.

Three months before the eclipse Aleksandr Ilych Ulyanov, a 21-year-old revolutionary, was executed for conspiring to assassinate the Tzar on the sixth anniversary of his father's death.

Bad move. Blood vengeance was to be served cold. Thirty years later his younger brother, Vladimir Illych Lenin, would seize power through the Bolshevik party and facilitate the extinction of the Romanov dynasty.

Time would tell that only a master intriguer and bank robber such as Stalin could ride Russia's greasy governmental tiger as Tsarists, revolutionaries, Mensheviks and Bolsheviks found, often paying the cost of attempting to do so with their lives.

Arthur Koestler identified the eclipsed nature of Russian politics in his book entitled Darkness at Noon as did George Orwell in his satirical expose of Utopianism, Animal Farm.

Apart from the temper of these three nations there were direct territorial implications which flowed from this eclipse which found its more immediate expression in the weakest link along its line - Manchuria.

All three eclipsed countries were to focus on acquiring the lands of Northern China in the coming years.

Russia was to start on the Trans-Siberian railway following the line of the eclipse very closely a few years later in 1891. Already controlling Outer Manchuria, it continually pressed for more influence in the inner half of the province through its rail lines.

Germany was to take over the Shandong peninsular within a decade with Japan starting to making inroads on the Chinese mainland a few years earlier.

Japan was to stop Russia in its tracks in their war of 1905 and to go on to take over both Manchuria and Shandong for itself.

This eclipse's demons were exorcised from Berlin and Tokyo in 1945 when they were occupied by the Allies at the end of World War 2.

Moscow, re-inforced by the July, 1945, Cold War eclipse still had to be contained by the USA and its Western allies until the Soviet collapse in 1990. It has never been defeated in the same way the other two were and not all its demons from this eclipse have been quelled, even today.

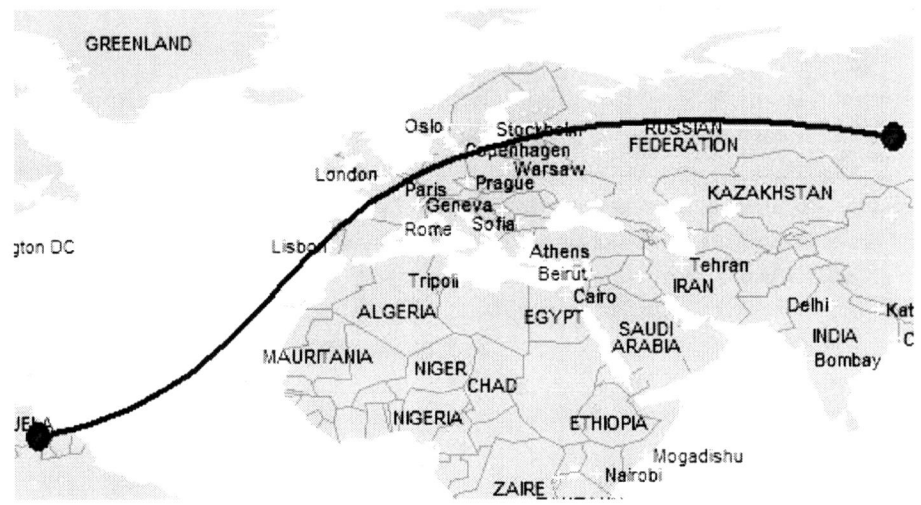

First World War Western Front Solar Eclipse of April 17, 1912

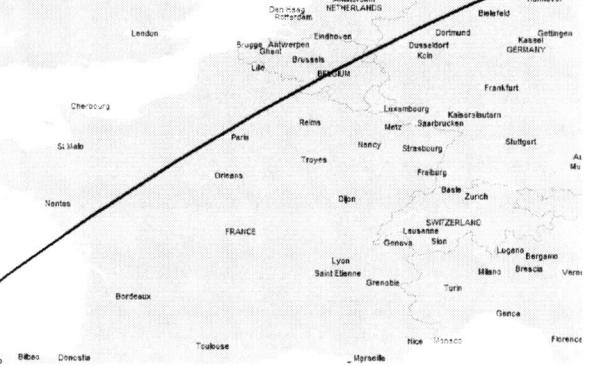

Occurring three days after the shocking sinking of the Titanic ocean liner this eclipse passed directly over what would become the graveyard of many young French, British, Commonwealth, American and German soldiers less than 30 months later.

The symbolism of the Titanic is still with us today and in retrospect the sudden and unexpected plunging into the icy abyss of the North Atlantic of the luxury liner is an apt metaphor for the switch from idyllic Edwardian aristocratic and village life to the explosive termination of a journey's end in the trenches at the Western Front.

This eclipse traced the southern arc of the path taken by the German Army through Belgium in 1914, the massacres of civilians at Liege, Leuven and Dinant located immediately either side of it.

Passing just north of Paris it outlined the southernmost bulge of the German conquests in France in 1917 where much heavy defensive fighting took place.

The eclipse also passed directly over the Ruhr, Germany's manufacturing workshop, including Essen, the site of the Krupp armaments factories and Europe's greatest industrial enterprise at that time.

The Krupp factories were crucial to the Kaiser's and Prussian martial confidence as they would be to Hitler's two decades later. As the blacksmiths of German nationalism Krupp had provided the weapons that brought Prussia the victories over Denmark, Austria and France in the mid-nineteenth century that facilitated the formation of the German State in 1870.

This eclipse also went through Prussia, the heartland of German militarism, no doubt encouraging the Junkers, its officer, land-owning class, who had hitherto experienced such a run of success, to believe another war would bring them yet more spoils.

Their martial approach to politics, which found its way into the writings On War of one of their number, Carl von Clausewitz, was exemplified by the Krupp-made heavy siege guns with near 20-foot barrels somewhat contradictorily-named Big Berthas. These culminated in the 100-foot long Kaiser Gun, capable of shelling Paris from over 80 miles away.

Whilst the Krupp factories were to become very busy and no doubt profitable during the total war that followed, defeat was incurred, in part, because they could not produce as many shells as the Allies.

To make a comparison with the devastating explosion of the Indonesian volcano, Krakatoa, in 1883 the eclipse over Essen created a "Kruppatoa" of industrialised murder.

This eclipse line's proximity to Berlin and Paris indicated the especially high losses of French and German soldiery that would follow.

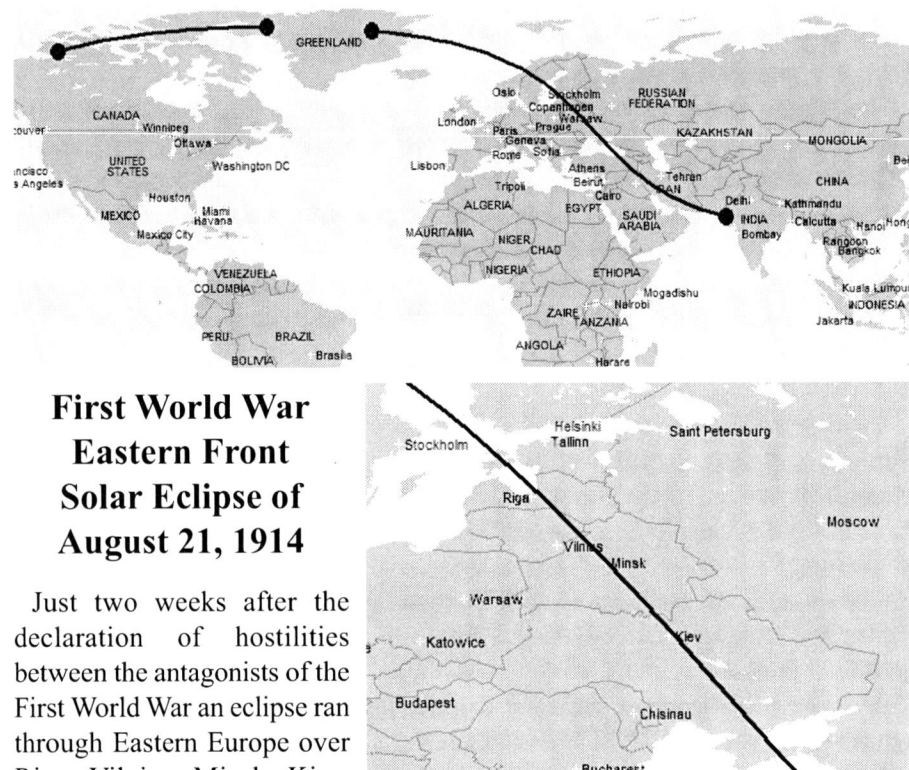

First World War Eastern Front Solar Eclipse of August 21, 1914

Just two weeks after the declaration of hostilities between the antagonists of the First World War an eclipse ran through Eastern Europe over Riga, Vilnius, Minsk, Kiev, the Crimea and down through the Black Sea to the Eastern Ottoman Empire and on to Iran.

It was to indicate the locations where some of the most terrible fighting was to occur during the period of the First and Second World Wars. It marked the additional conflicts around the establishment of Bolshevism, the White Russian counter revolution and the destruction of the Kulaks, Russia's independent farmers. Imperial rivalries were complicated by nationalism, class war, ethnic segregation, historic territorial claims and foreign intervention.

This incendiary mix of competing objectives led to protracted resolutions and constant re-openings of bitter conflicts.

Riga, the capital of Latvia, was part of the Russian Empire at the time of the eclipse. Within a year it was on the frontline, a fierce naval battle taking place between German, Russian and British gunships and submarines in the Gulf of Riga directly under the eclipse line.

Latvian army units were formed to fend off the German land advance with very high casualties. Its industry was dismantled and taken to the Russian rear and its agriculture destroyed in a two-year stand-off during which Riga held out despite the German conquest of much of the countryside.

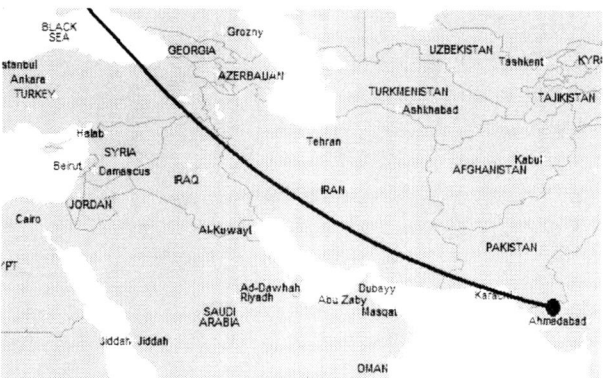

In September, 1917, the Germans entered the port city. The occupation lasted just over a year to be followed by Allied recognition of an independent Latvian state, contrary to the wishes of the Soviets who wanted it to return to Russian control.

Latvia became embroiled in the ensuing Russian civil war between Reds and Whites, entailing two more years of fighting between the Bolsheviks, Allies, Latvian nationalists and the Baltic Germans known as Balts.

One can only imagine the crises of identity and decisions about which side to take that went on. Were you a Balt, a Russian or a Lat? Were you a nationalist or an imperialist, a Soviet or a Republican? Did you keep your head down or put it above the parapet? Did you look after your family or your political community? Such is the confusion and hard decision-making that eclipses can require. In 1920 the Allies left and Latvia signed a treaty with the Bolsheviks securing its independence.

Vilnius, the capital of Lithuania, was also a part of Greater Russia at the time of the eclipse but its population was more Polish and Jewish than German or Russian.

It was captured by the Germans by September, 1915 and then caught between the Soviets and a resurgent Poland after 1918, ending in Polish annexation in 1920. Rural Lithuania did achieve independence though, albeit under autocratic rule following a military coup d'etat. World War II saw Lithuania taken over first by the Soviets who murdered the Balts followed by the Nazis who murdered the Jews.

The people of Minsk were equally unfortunate. The city became a battle-front, hospital town, during 1915, when it was selected as the headquarters of Russia's Western Army. It then became a centre for the Bolshevik Revolution in 1917 but was taken over by the Germans as part of the Treaty of Brest-Litovsk. It then exchanged hands between Byelorussian nationalists, Soviets and the Poles until eventually settling down as a Soviet Republic in 1920. When the Germans invaded in 1941 Minsk fell quickly and was requisitioned, leading to the starvation of its inhabitants. Heavy fighting took place before the Soviets regained it in 1944, some four fifths of the city being destroyed in the process and its population being reduced by more than 80%.

In comparison Kiev was relatively well off during the First World War, only being taken by the Germans two days before the March, 1918, signing of the Treaty of Brest-Litovsk.

But its luck ran out and in the aftermath of the War it changed hands 16 times in less than two years in the battles between Red and White Russians, Poles and Ukrainian nationalists.

It became a Soviet Republic in August, 1920. It was to suffer the Great Famine in 1932 and 1933 and the Great Purge of 1937-38 before being occupied by the Nazis in 1941. The retreating Red Army detonated 10,000 mines before they left destroying most of the city's buildings and causing a fire storm lasting five days. Babi Yar, a Kiev suburb, was the site of a massacre of nearly 34,000 Jews.

The End of the Jewish Pale

This Eastern Front eclipse cut right through the middle of the Jewish Pale, an extensive area of Western Russia granted by Catherine the Great in the late 18th Century to the Diaspora.

At the time of the eclipse some five million Jews, half their total number, were confined to living there, many subject to injustice and harassments known as pogroms.

From comprising up to 15% of the population of the region, barely any live there now, having been scattered during the First World War and rounded up and killed in the Holocaust in the Second.

The Pale was formally abolished in 1917 when the Bolsheviks came to power.

Perhaps this eclipse was required for the Jews to be given additional impetus for the establishment of Israel?

The similarity of the eclipse line to the 1918 Treaty of Brest-Litovsk boundary between Germany and Austria-Hungary and the pacifist Bolshevik Russian government is particularly striking.

It can also be coupled with Hitler's idea of Lebensraum, literally the "living space" that Germany felt should take its national boundary in the East at least to the line of the eclipse. This Nazi concept was at the heart of its inhuman treatment of many of the peoples living under the eclipse's path contributing to a death toll said to be as high as 50 million.

This litany of war-zone horrors was not confined to the eclipse's path North of the Black Sea. It continued to the South of this water as well.

Where it passed through Eastern Turkey became the sites of the Armenian massacres in 1915 in which the Turkish government systematically slaughtered the men and forced marched the women and children into deserts.

The methods of extermination were completely indiscriminate and heartless. They included burning, drowning, poisoning, gassing and drug overdosing. An estimated 1.5 million victims died in these ways prompting the use of the term genocide for the first time.

The eclipse also went through Iran, which, despite, trying to maintain its neutrality was swept up in the conflict. British, Russian and Turkish armies fought on Iranian soil during WW1 and the 1919 treaty with Britain turned it into a virtual Protectorate.

This eclipse can be linked to the fall of five dynasties, one of which, the Hapsburg, went back 700 years, passing either directly through their lands or to the side of them. The earlier Western Front eclipse also passed through two of them re-inforcing their absolute fall from power.

Those that were removed within a few years included the Romanov Emperor of Russia, the Hohenzollern Emperor of Germany, the Hapsburg Emperor of Austria-Hungary, the Osman ruler of the Ottoman Empire and the Qajar Emperor of Iran.

In particular the lands of the Austro Hungarians and Ottomans were re-organised into nation states in a grand shuffling of the pack, the consequences are with us today, having been played out in the Balkan Wars of the 1990s and the present Syrian Civil War.

It may or may not be significant, given the difference in time, but this eclipse also passed over Chernobyl which is situated in the Ukraine north of Kiev. Seventy three years later it was the site of a nuclear re-actor which went into meltdown causing many deaths from radiation and the evacuation of the land area around it.

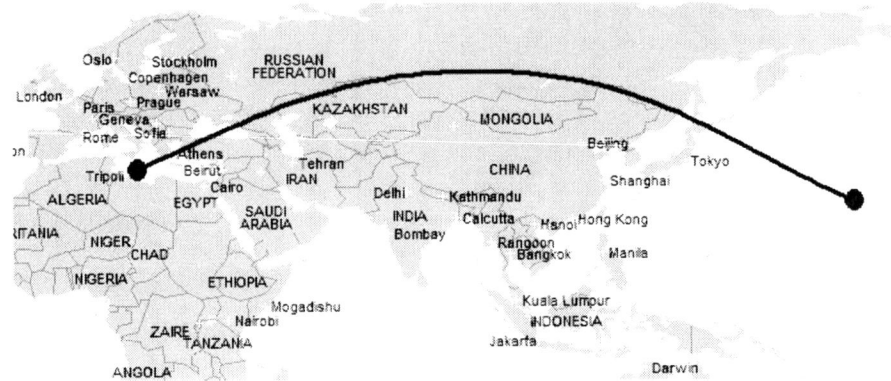

Second World War
Solar Eclipse of June 19, 1936

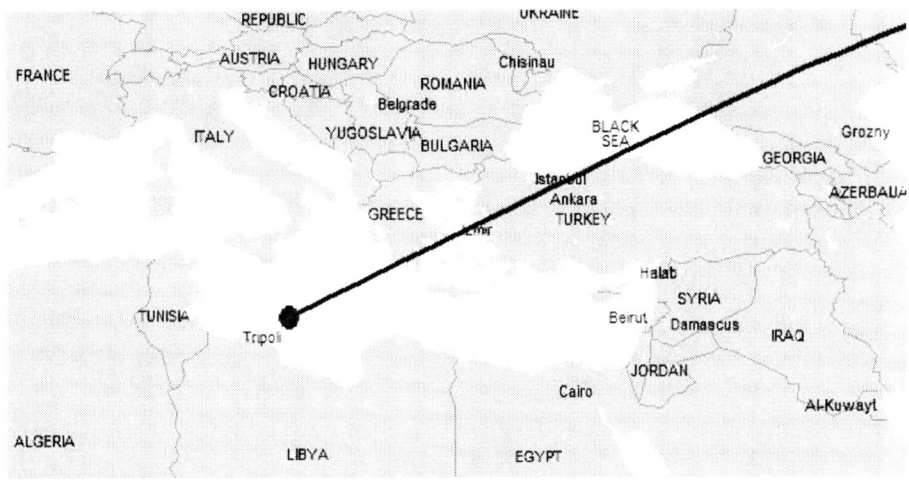

Understandably most Anglo historians hold 1939 as the year that the Second World War started in Europe and 1941 for the Pacific. This eclipse indicates it started earlier.

Hitler was already on a war footing in June, 1936, having remilitarised the Rhineland three months before in order to prepare for his attacks that came in subsequent years.

The Allies were just not ready to accept that they would have to fight him as he took back the Saarland in 1935, marched into Austria in March, 1938 and absorbed the Sudetenland of Czechoslovakia six months later.

This eclipse does not repeat the European war zones of the two First World

War eclipses which were to be re-activated by the fresh conflict but rather opens up the southern Mediterranean front which was to see fighting in new areas including Greece, Italy, 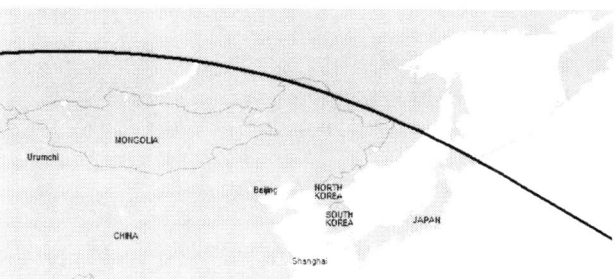 Crete, Malta and North Africa.

This eclipse began between Italy and its colony, Libya, spelling disaster for the imperial power as it was directly cut off from its overseas lands.

North of the eclipse line Germany was to be initially successful in Greece, albeit it helping out its failed Italian allies.

Arguably the war in the southern front, which saw such battles as Tobruk and El Alamein, was won by the British Navy which destroyed the German supply lines to its North Afrika corps so it is fitting that this eclipse started over water.

Libya, which is the base from which the Italians began the North African campaign and through which the Germans were supplied, is also highlighted.

A month after this eclipse the Spanish Civil War began. It missed the Iberian peninsular suggesting that Spain and Portugal would not be combatants in the wider theatre, although the Nazi support for the Franco fascist rebellion, indicates that it was part of the general conflagration.

Extending the line of this eclipse westwards takes it towards Spain's then African and Atlantic lands from which Franco launched his attack on the new Spanish Republic and where he had learned the savage treatment of colonial peoples that he was to turn on his own people.

It also went over the Caucasus. When the Germans invaded Russia its armies were unexpectedly diverted to this region away from the more obvious target of Moscow. Ostensibly it was because the oil fields in the region but perhaps the magnetic attraction of this eclipse had something to do with it.

Stretching deep into Russia this eclipse not only revealed the heavy losses the Soviets would incur but also indicated that the whole of the country would have to be brought into play in the conflict.

In the East, Japan was to invade China a year after this eclipse whose line encircled both combatants to their north. The invasion of China was to see dreadful treatment of the Chinese who experienced mass flight from devastating aerial bombardments and sadistic cruelties perpetrated upon both prisoners and civil populations, the Rape of Nanking amongst them.

An estimated 20 to 30 million died in the conflict which paved the way for Communist Party and Red Army rulership of the country. This eclipse fizzled out in the Pacific where much of the second phase of the fighting was to take place.

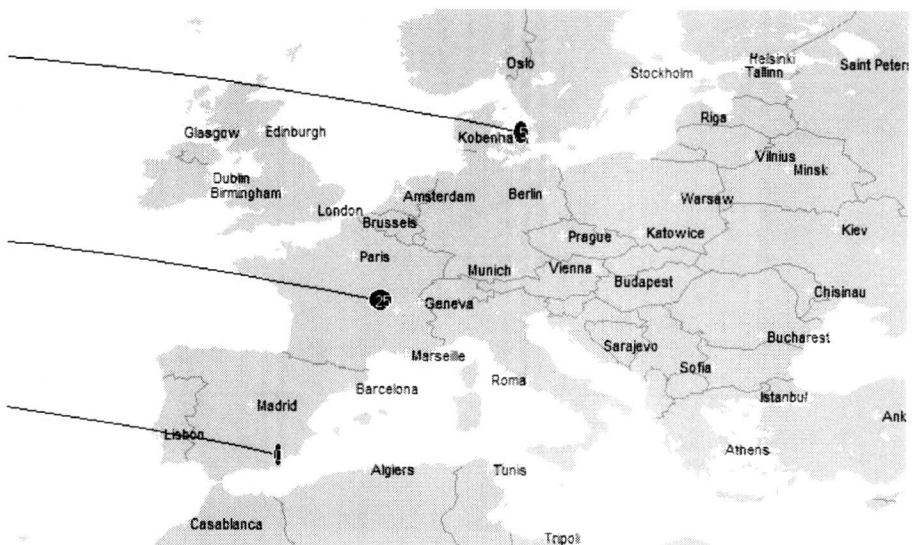

World War Two Western Front Partial Eclipse of April 19, 1939

Just over a year after this partial eclipse over them, France, Britain, Belgium, Denmark, Norway and Holland were attacked by Germany's Panzer tanks in fast-moving Blitzkriegs.

Being only a partial eclipse it meant not all would be lost but it was not far from it.

All bar Britain were forced into occupation and it only survived by the skin of its teeth managing to withdraw much of its army from the French port of Dunkirk and proving able to withstand the aerial contest with the Luftwaffe, the Battle of Britain.

If it had been a full eclipse then Britain would not have been able to retain its independence. As it was it could fight another day.

Sweden was barely touched by the eclipse and was thereby able to remain neutral during the conflict. So too with Switzerland.

The civil war in Spain, from where it was only slightly visible, ended just two weeks before the eclipse whilst the southern lands of Vichy France remained unoccupied, probably because, from there, the eclipse was less than 25 per cent visible.

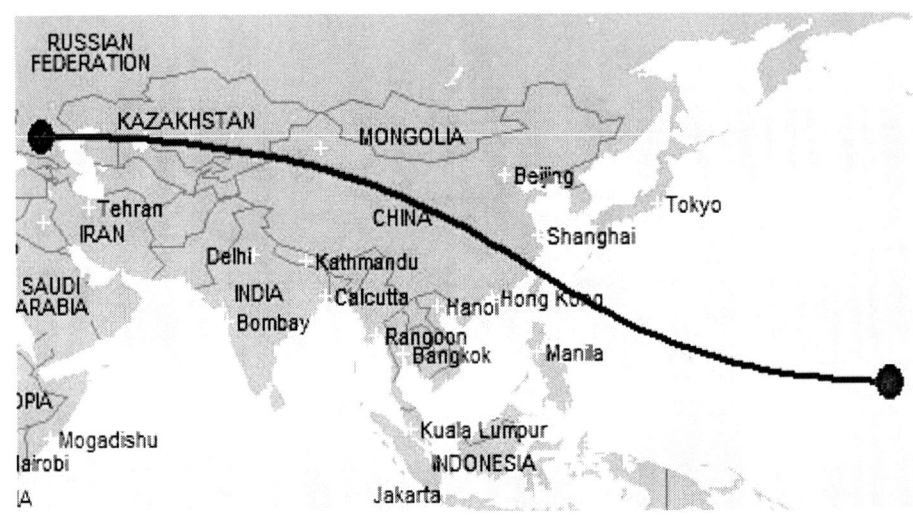

Second World War Victory Locations Solar Eclipse of September 21, 1941

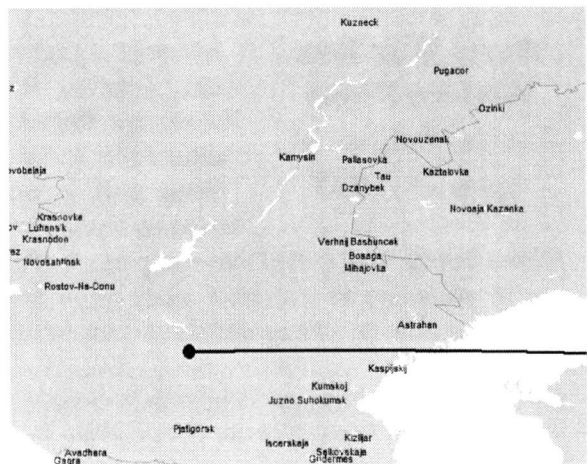

This eclipse occurred three months after Germany descended upon the Soviet Union and three months before the Japanese attacked Hawaii's Pearl Harbour naval base.

Although timewise it occupied the halfway point between the starts of the major battles in the European and Pacific theatres it is particularly interesting spatially as it indicated the locations where the victories were going to be achieved.

The eclipse started in the North Caucasus a short hop from Stalingrad now renamed Volgograd. It was here that Germany's armies were stopped and surrounded, effectively confirming that the Allies were going to triumph in the West.

The eclipse also passed over Guam in the Pacific. An outpost of the USA before the War, when it was recaptured from the Japanese it enabled bombing raids on the Japanese mainland itself including firebomb attacks on its capital, Tokyo.

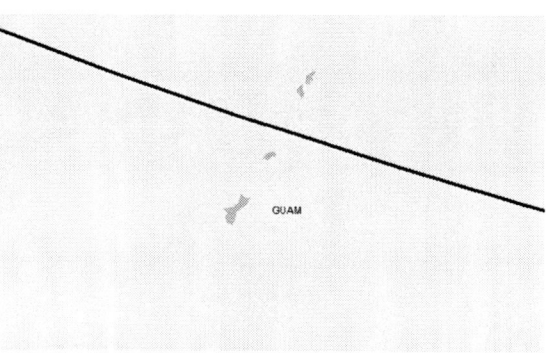

From Tintao, a small island just above Guam, the B59s carrying the atomic bombs that fell on Hiroshima and Nagasaki, took off and landed. Subsequently the Emperor of Japan raised the White Flag.

The Japanese Empire was destroyed from this advance Pacific base, marked out by this eclipse, ushering in the Age of Mutually Assured Destruction (MAD) by nuclear weapons.

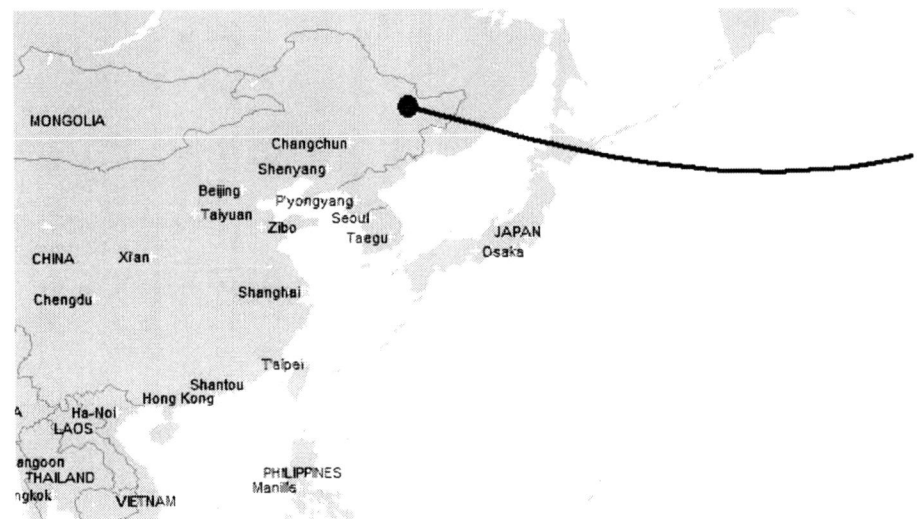

Japan to Lose World War Two
Solar Eclipse of February 4, 1943

Five days after this eclipse, visible from all of its mainland, Japan's Imperial army withdrew from Guadalcanal ending five months of rearguard fighting. They were now on the back foot on land as well as sea as the Allies began to claw back from them the impressive gains they had made during 1942. The days of overrunning everything in their path were finished and they were to pay grievously for their ambitions.

Having added Indochina, Burma and the Philippines to their conquests and even attacked Australia and the Aleutian islands the Japanese were on a downward slope after this eclipse over their homeland, culminating in the bombing and fire-bombing of all their major cities and the use of nuclear weapons on a civilian population.

This eclipse also signalled that the end of the Japanese as a military power was not too far away, their samurai swords to be either melted down to make (figuratively) cars and computers or to be taken back to the USA as trophies in the backpacks of the victorious GIs.

This hugely proud and in many ways meritorious nation had mistimed its challenge to the USA after its humiliation by Commander Perry 90 years earlier. Despite its frenetic drive for modernisation its geographic location, population and resources meant its place in the world could never be number one. Commander Perry had wanted them merely to trade in a friendly way and that is what would now be emphatically enforced upon them.

African, Indian and South East Asian Independence
Solar Eclipse of July 20, 1944

A year before the end of the Second World War this eclipse raised the issue of the post war settlement in European colonies in South and South East Asia. It started in the Horn of Africa and travelled through the Indian subcontinent, Indochina, the Philippines and Indonesia.

Whilst it was clear at this stage that it was only a matter of time before the Allies were going to defeat the Axis powers the initial success of the Japanese in overrunning European colonies emboldened those indigenous nationalists seeking independence.

Savage conflicts were to erupt in the whole region as local peoples fought off imperial attempts to return to the pre-war status quo and establish their new national potencies.

India and Pakistan achieved independence three years later amidst blood and displacement, Burma following four years later. Vietnam was to undergo over thirty years of civil war and Laos and Cambodia to suffer their own tragic ravages.

Each country fared differently depending on local circumstances but the die was cast and national self-determination in the regions passed over by this eclipse would be achieved.

This was most clearly stated six months after the eclipse when the Anglo-Ethiopian agreement was signed whereby Britain, who had done much to end Italian control of the African country, lost many of its advantages in the Horn of Africa country including its precedence over other foreign representatives.

Ethiopia was referred to as an "ally" in the agreement, thereby granting it equal national status and a seat at future peace conferences.

Although insufficiently established to be a general rule it can be noted around this eclipse that, at its inception in the Horn of Africa independence was given freely by the imperialist sponsor; at its early midpoint in India it was given peacefully but received with violence; at its later midpoint in Vietnam there was prolonged war, first with the returning Imperialist power, then with itself and finally with an outside power; at its end the Philippines was granted a violence-free but strings-attached independence two years later in which they had to allow American bases on their territory as well as not compete economically. Indonesia's independence was to be marred by fighting with its former Dutch masters.

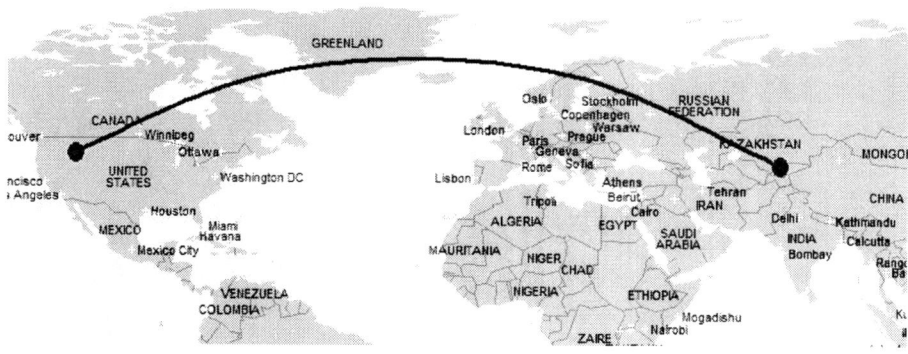

Cold War Eclipse of July 9, 1945

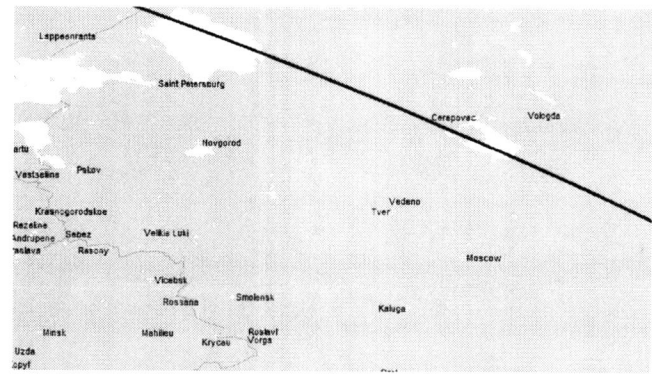

This eclipse ran from North West USA over Canada, the (cold) Arctic and down into Russia not too far from its capital, Moscow just two months after the end of the Second World War in Europe.

In doing so it marked out the world's two major antagonists for the next 45 years.

Just as it occurred the victors fell out. Stalin instructed the Red Army not to fraternise with the Allies and an Iron Curtain began to descend on Eastern Europe.

Fascism had been crushed. Now it was the turn of communism and capitalism to square off against each other. The fact that Washington was further away from the line of the eclipse than Moscow indicated that the USA would have the upper hand in this contest.

The fall-out from this epoch-making eclipse is still with us today in the suspicious relationship that continues between the USA and Russia.

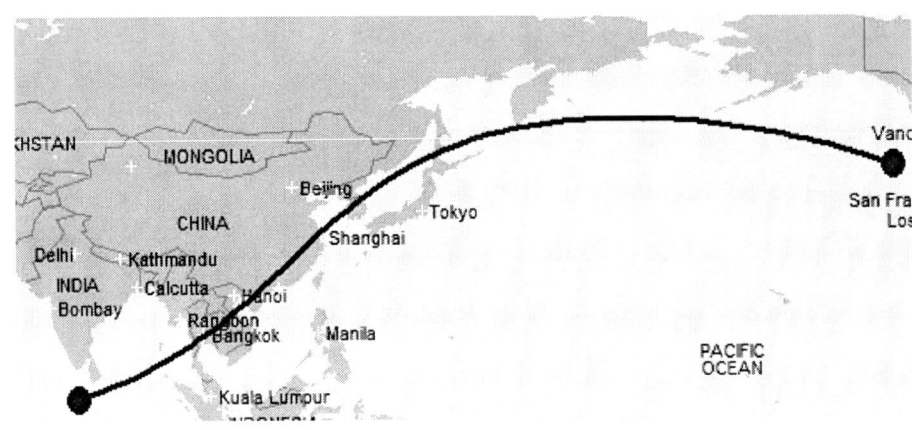

Korean War Solar Eclipse of May 9, 1948

Within four months of this eclipse bisecting the Korean peninsular the States of North and South Korea were formally established. Just over two years after this eclipse the divided halves were at war.

On the ground the former Japanese imperial colony was split into North and South parts along the 38th Parallel just above Seoul but in the heavens the division had been made further east of the new capital city.

This expedient settlement between the unnatural allies of the capitalist US and the Communist Russians cut across, not just the peninsular, but also Korean nationalism. These tensions turned septic when North Korea invaded the South on June 25, 1950 with a force, 135,000 strong.

Although the North captured Seoul and much of the peninsular, the US-led

United Nations intervention in support of the South turned back the invasion. At one time - January, 1951 - the boundary between the two warring states followed much of the exact line of the eclipse.

The Allies went on to capture Pyongyang, the capital of North Korea, prompting the Chinese to throw their weight behind the North leading to a stalemate and the re-instatement of the 38th parallel as the boundary between the two countries in 1953.

Technically the countries are still at war, the man-made boundary still in friction with the heavenly one and, arguably, re-actived by the May 21, 2012, eclipse which has seen intense threats of war between the two halves again.

Indochina/Vietnam Independence Wars Solar Eclipses of July 20, 1944 May 9, 1948 June 20, 1955 December 14, 1955 April 15, 1958 November 23, 1965

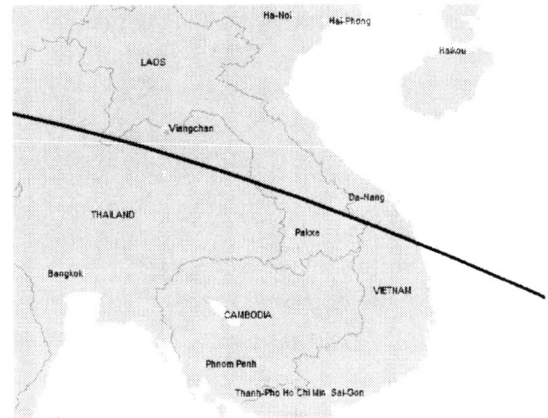

The first of the six eclipses which made Vietnam a war-torn country for 30 years was the already mentioned decolonisation eclipse of July 20, 1944, which swept from the Horn of Africa via India to Indochina.

Within 12 months the Viet Minh seized power under Ho Chi Minh, leading to the beginning of a vicious fight with France for colonial independence.

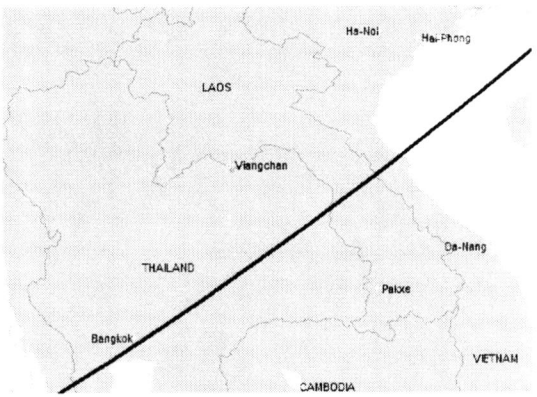

By 1950, after the second eclipse, the USA was supporting France and the newly Communist China was backing the Vietnamese nationalists culminating in the defeat of the French at Dien Bien Phu in 1954.

The two eclipses of 1955 co-incided with the division of Vietnam into North and South parts along the 17th Parallel

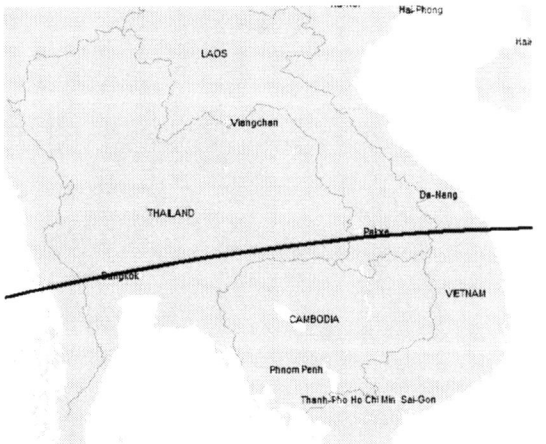

34

and the French withdrawal in 1956.

By the time of the fifth eclipse in 1958 North Vietnam was waging a guerilla war on the South and the Ho Chi Minh Trail through Laos and Cambodia to deliver weapons to the insurgents was up and running.

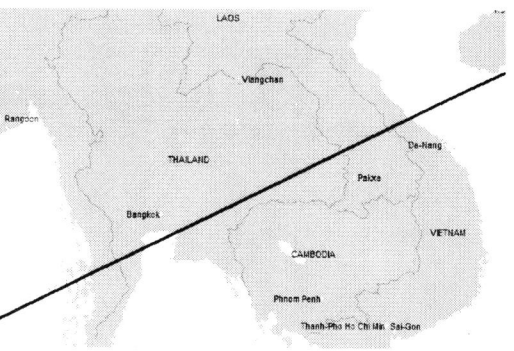

The conflict escalated until it became an all-out war around the time of the sixth eclipse in 1965 when the Americans committed 200,000 troops and began bombing raids on North Vietnam.

Two of the eclipses - in December 1955 and April, 1958 - went exactly over the demilitarised border which separated North and South Vietnam. The first was at the time of its establishment and the second occurred when it became a place of active conflict.

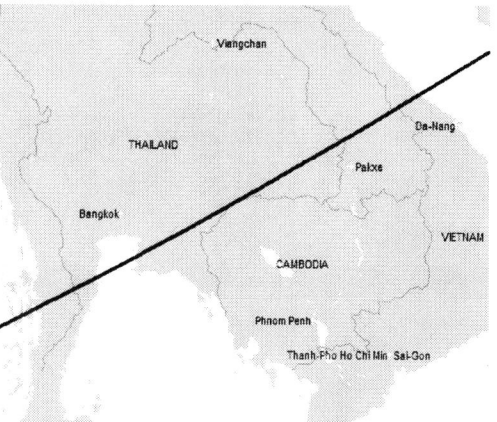

Atrocities were carried out by both sides there including those at Hue, an historic city close to the DMZ, by the Vietcong.

The June 20, 1955, eclipse went directly over the small village of My Lai, some 100 kilometres south of Da Nang.

It was here that the Americans, who always had a military advantage, were to lose the war through an irrational massacre of 500 Vietnamese villagers.

American public opinion,

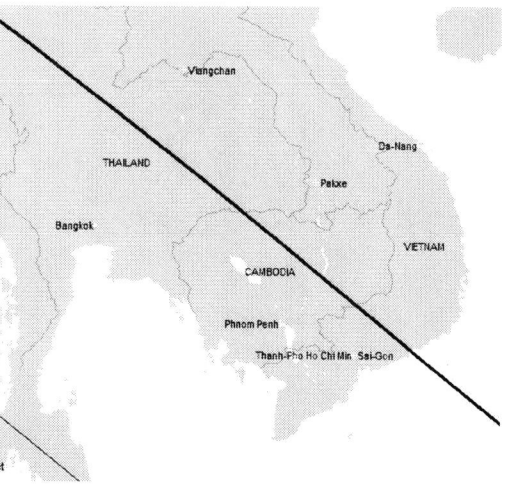

outraged by this loss of moral superiority, turned against the war, leading to withdrawal and defeat.

The eventual collapse of the South was also indicated by the proximity of the final 1965 eclipse to its capital Saigon.

Only the sixth of these eclipses passed directly through Cambodia but what followed was even worse than the fate which befell the Vietnamese.

Implicated by the establishment of the Ho Chi Minh Trail on its border with South Vietnam the Americans bombed it as well leading to the establishment of a murderous Communist regime under the dictator Pol Pot who systematically wiped out up to two million of his countrymen, many of them urban intellectuals who were made to work as peasants.

NATO - Islamist Wars
Solar Eclipse of August 11, 1999

Starting in the North Atlantic this eclipse linked the two antagonists of the Western democracies and the Al Qaeda Jihadists and despots of the Islamic world.

It is an era eclipse, superceding the Cold War period and pitting the high-tech forces of the North Atlantic Treaty Organisation of North America and Europe with fanatical terrorist groups and rogue states.

This eclipse's energies, compounded by a

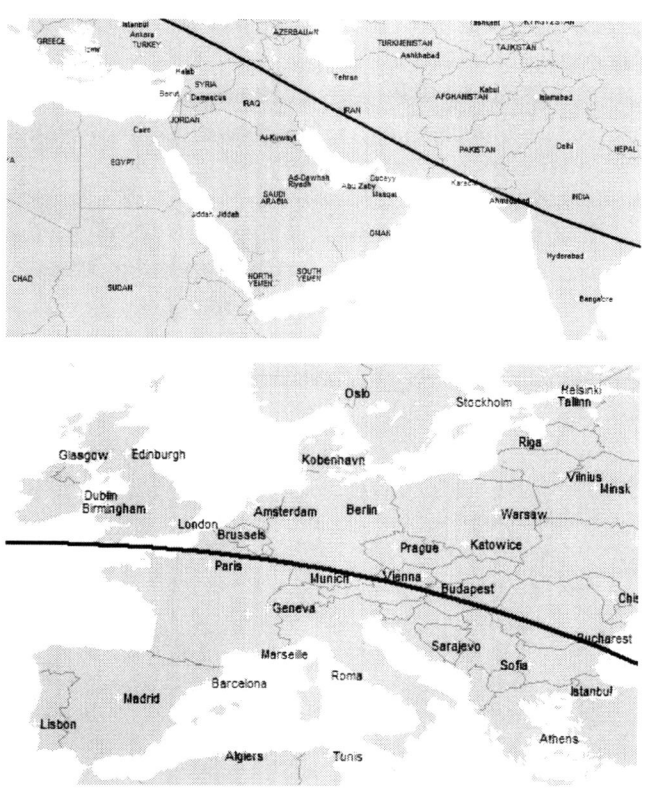

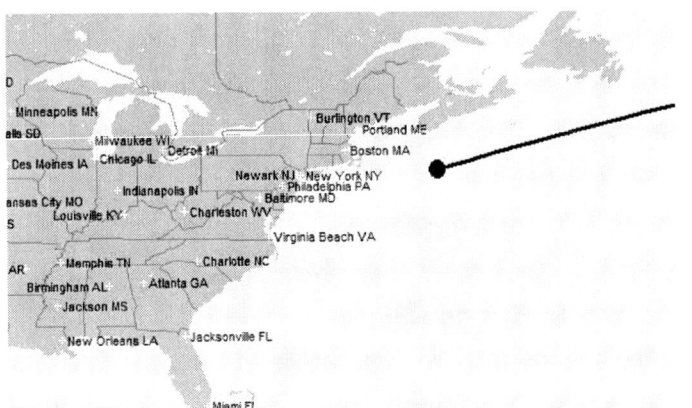

later partial eclipse over North America the following year, found its way into the September 11, 2001, aircraft suicide attacks on the Twin Towers of New York and the Pentagon in Washington. This atrocity was to lead to the subsequent invasions of Afghanistan and Iraq by the USA and its allies. All four parties to these conflicts were linked by this eclipse.

Its line did not reach the US mainland but it lay offshore of New York and Washington and may be classified as a pointer eclipse as well as an indicator of regional combatants.

In Europe it sliced through Cornwall, the South West tip of the UK, heralding the 7/7 London transport bombings and also went over Northern France and Southern Germany suggesting those countries will have serious issues to face from Muslim extremists.

Cutting diagonally through Iran and through the south of Pakistan it exacerbated the tensions of those two countries both of which are at loggerheads with the USA and feeding Islamic terrorism.

Whilst it went directly overhead of the Kurdish lands of Northeast Iraq, Afghanistan was just to the north of the eclipse line, experiencing a 90% occlusion.

Christians are becoming thinner on the ground in the Middle East as militant Islamists seek to dominate the region and the two religions become more polarised.

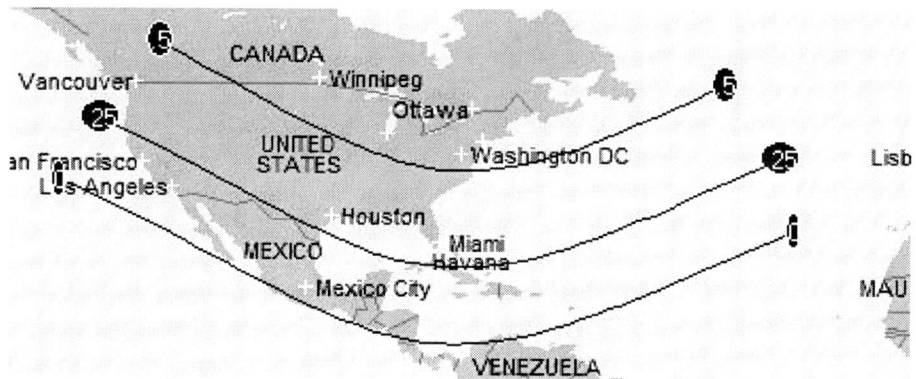

World Trade Centre Attacks
Partial Eclipse of December 25, 2000

Whilst the 1999 total eclipse outlined the new geopolitics this partial eclipse instigated what must be considered to be one of the world's most subliminal, spectacular and distressing events in human history.

The shocking eradication of the Twin Towers of the World Trade Centre by ramming them with hijacked passenger aircraft stunned the world and plunged Americans into a deep national mourning.

The event brought about an abrupt awareness that they faced sinister and ruthless opposition in significant sections of the globe.

Freedoms were lost, suspicions grew about foreigners residing in the USA and America's armed forces were let loose on specific target countries in the Middle East and South Asia which either harboured al-Qaeda operatives or could threaten the use of weapons of mass destruction.

Overall the world appeared a much less safe place after this terrorist event which occurred nine months after the eclipse. Fortunately the eclipse was only partial - around 60% visible over New York - or it could have been much worse.

Although several thousand American lives were violently cut short in the attack on domestic soil many times more lives were to be ended in the wars that followed in Iraq and Afghanistan both situated along the line of the total eclipse two years earlier.

Another reflection of its partiality was that one of the four hijacked planes was downed before it reached its objective in Washington.

The USA's 'Manifest Destiny'

How 19th century North and Central American eclipses marked the expansion of the USA's 13 colonies through wars, acquisitions and boundary confirmations to become a 50-state transcontinental nation and a global power.

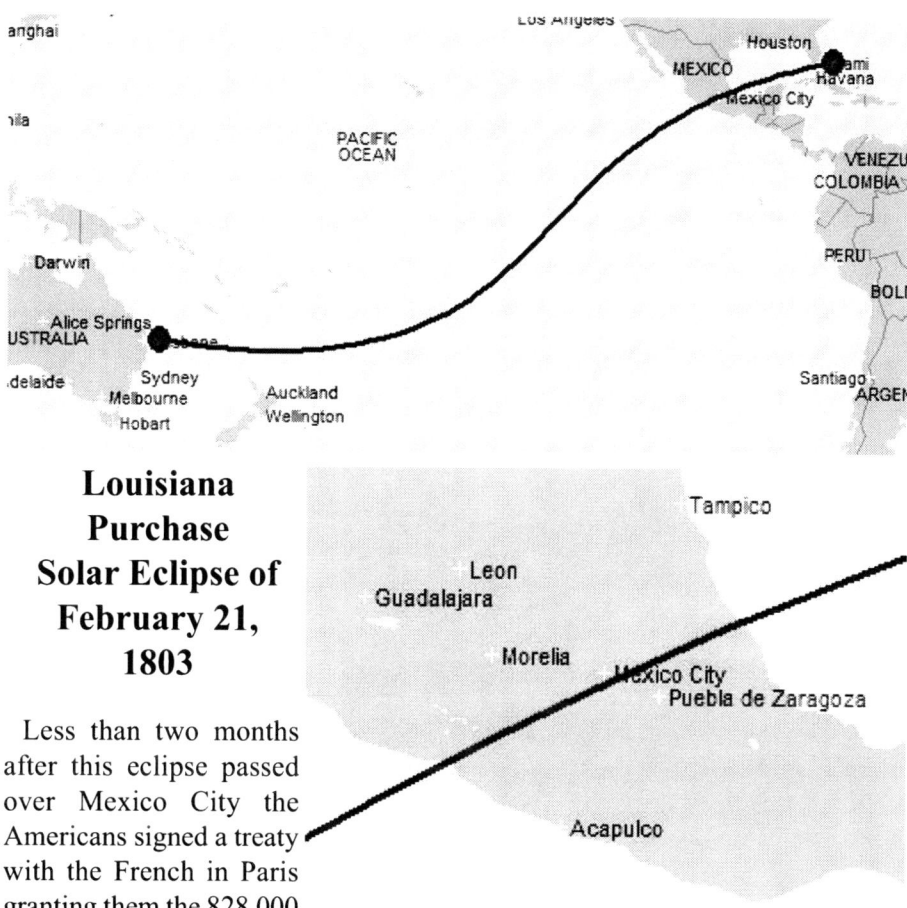

Louisiana Purchase Solar Eclipse of February 21, 1803

Less than two months after this eclipse passed over Mexico City the Americans signed a treaty with the French in Paris granting them the 828,000 square miles of Louisiana, France's North American lands based around the Mississippi river.

One of the signatories, Robert Livingstone, observed of the acquisition: "From this day the United States take their place among the powers of the first rank."

Hitherto Spain had dominated the vast majority of the land area of the Americas. Indeed for a good part of the eighteenth century it had also controlled these French lands, Napoleon only recovering them for France in 1800 after his conquest of Spain.

Eclipses are indicators of loss. They behave like snakes in the game of Snakes and Ladders. For Mexico the writing was on the wall. The Americans were the coming power in North America. The Spanish Mexicans were being eclipsed.

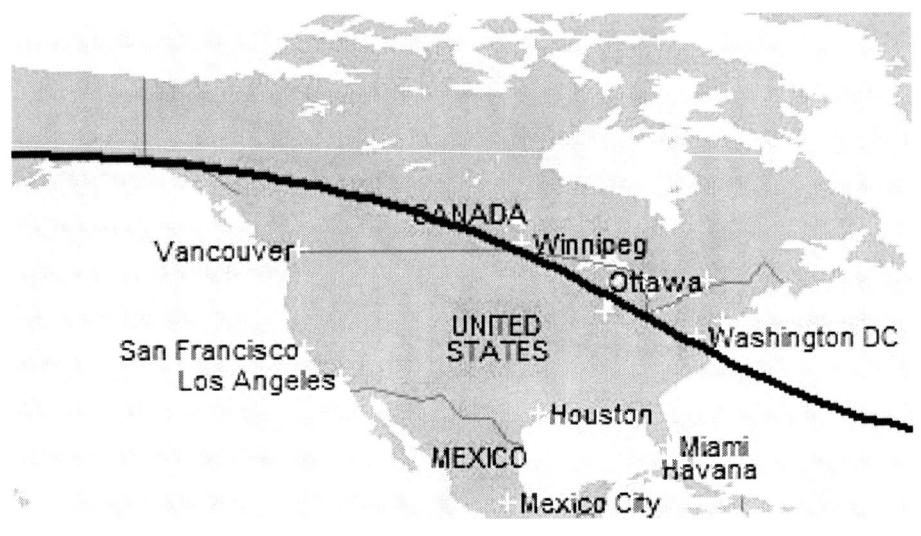

US-British War Solar Eclipse of Sept 17, 1811

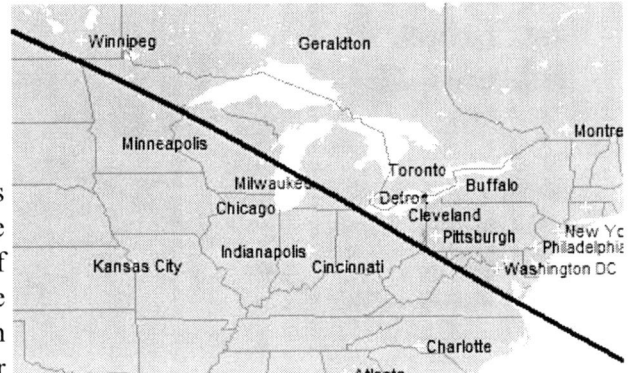

Ten months after this eclipse traversed the British territories of North America and the 13 States of the USA on the Eastern seaboard war broke out in a dispute over the boundaries between them.

It lasted three years and was fought in three main theatres - the Great Lakes, the Atlantic Coast and the Mississippi, two of which are directly crossed by the eclipse.

Although each others' territories were occupied and repulsed during the conflict the original boundaries were re-affirmed in the subsequent Treaty of Ghent.

The Americans had little success in those areas to the North of the Eclipse but won more convincingly in the conflicts to its South.

Although eclipses at sea are less likely to engender conflicts, in this case one of the major causes of the war was the upset in the USA over Britain's willingness to board American ships to search for sailors to man their vessels.

The end result was something of a stalemate, although it is surprising that the USA did as well as it did given that the eclipse passed close to its capital, Washington. However, in line with the usual difficulties associated when eclipses go overhead, the British did inflict on the Americans the indignity of burning down the White House.

Canada, which was now definitely not to be absorbed into the Manifest Destiny of the United States, still celebrates the anniversary of the war as a way of asserting its own nationhood.

The Northwest Territory to the South and West of the Great Lakes and comprising the States of Ohio, Indiana, Illinois, Michigan, Wisconsin and Minnesota were to fall indisputably under US control.

Although ceded to the fledgling revolutionary United States by the British in 1787, these lands were still to be properly occupied and had become centres of resistance by native American tribes supported by the British in a rearguard action.

Although clearly rebuffed the USA can be said to have moved up a further notch in the global pecking order, officially establishing itself as an independent challenger to British global power. Possibly the biggest losers of the conflict were the native American Indians of the Northwest Territory which was exactly eclipsed.

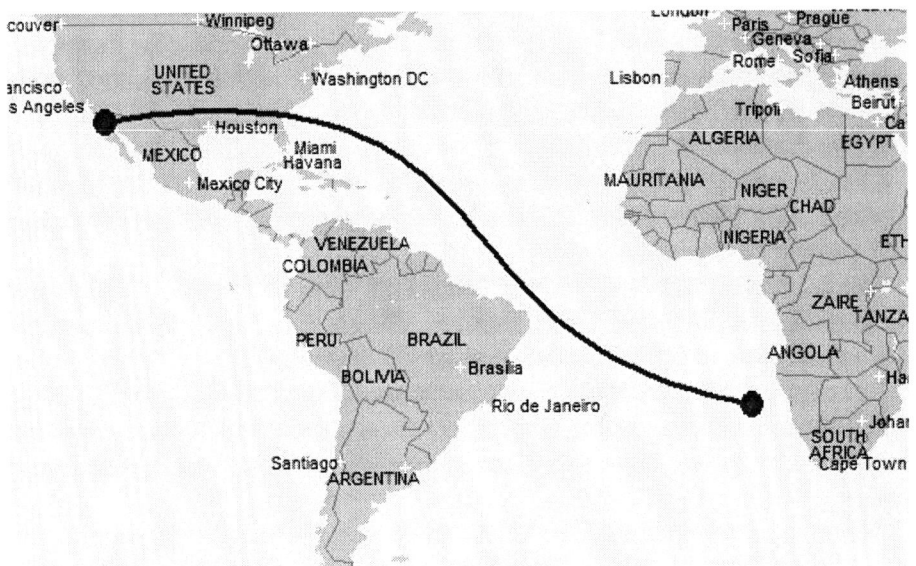

Monroe Doctrine (Mexican Independence) Solar Eclipse of September 21, 1821

Occurring just six days before Mexico formally achieved independence from Spain this eclipse divided this new country in two almost exactly along the line the USA was going to snatch its northern territories less than 30 years later.

Although it marked the conclusion of 11 years of struggle for independence from over 300 years of colonial plunder it indicated the fragility of the integrity of Mexico's newly acquired nationhood and its vulnerability to the predations of its longer established, better organised and better resourced north eastern neighbour.

The opportunity to dominate the Americas and sense of the new potential of the USA to challenge the European colonisers following Mexican independence was expressed most emphatically less than 30 months later with the pronouncement of the Monroe Doctrine on December 2, 1823.

This policy, articulated by US president James Monroe, stated that further efforts by European nations to colonise the Americas would not be tolerated.

This assertion would be insisted upon frequently in the forthcoming decades, reflecting the line of the eclipse into the South Atlantic, off the Brazilian Coast and across to offshore Africa literally cutting off Spain and Portugal from their former South American colonies.

44

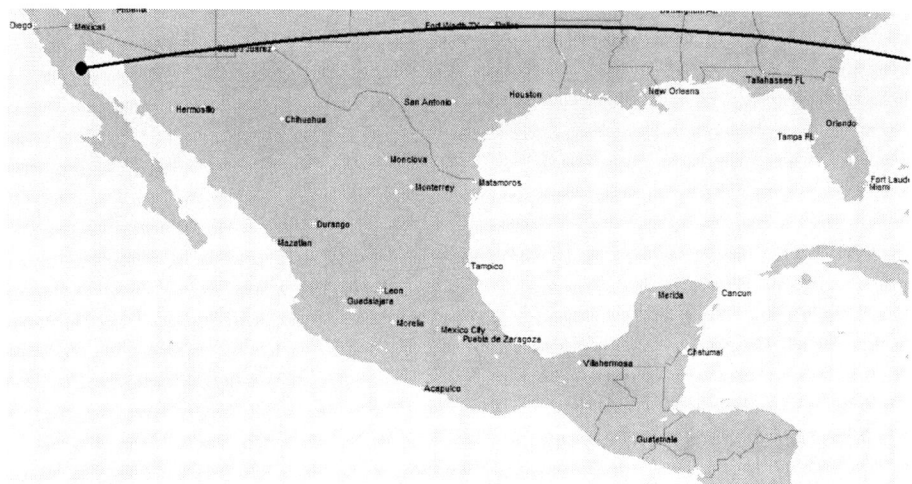

Although there are other more timely eclipses marking the later Texas and US wars with Mexico and the American Civil war this eclipse line fed into the tensions of these future conflicts.

Noticeably it passed through the American slave states where there was considerable ill-treatment of African Americans by plantation owners at this time.

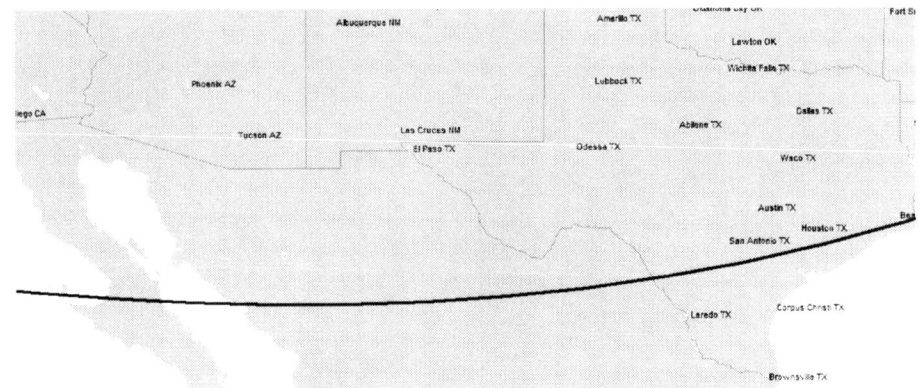

Texas Independence Solar Eclipses of February 12, 1831 and November 30, 1834

Ten months before the first of these eclipses, Mexico issued an edict warning that there should not be any more immigration from the United States into its then regional state of Texas.

But the pressure for Americans to continue to settle proved too great. Sixteen months after the eclipse, which passed through San Antonio, Austin and Houston, a Mexican army was defeated by Texan insurgents. The conflict that was to lead to first Texan independence and then integration into the USA had begun.

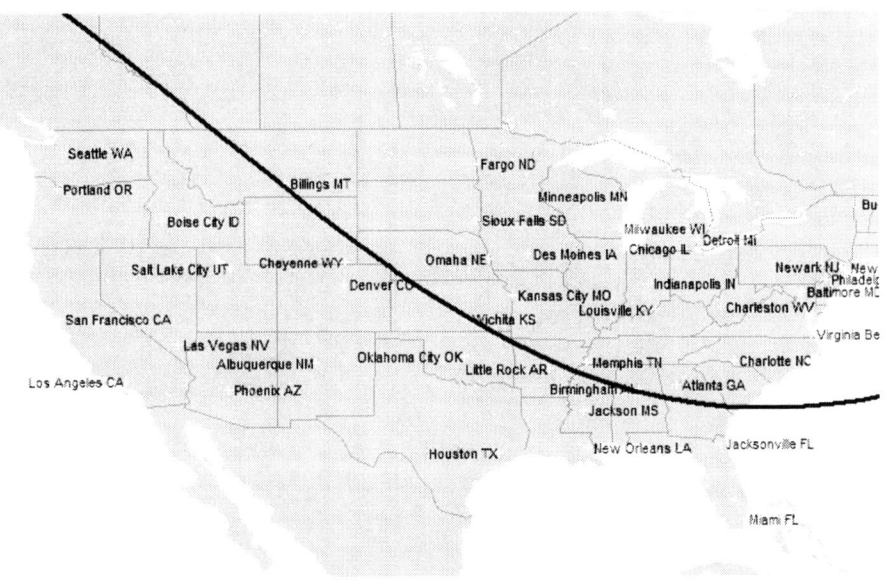

Less than a year after this second eclipse the first shot of the Texan revolution was fired. Within five months independence was proclaimed following the heroic defence of the Alamo and the subsequent defeat of the Mexican Army.

Although the eclipse was not total in Texas it did pass through Tennessee, 15 of whose citizens under the leadership of the frontiersman Davy Crockett turned up and fatally volunteered to fight on behalf of the Texan rebels at the Alamo.

Texas, however, was to remain independent of the USA for some ten years when its wished-for incorporation into the USA was one of the influences that prompted the US-Mexican war.

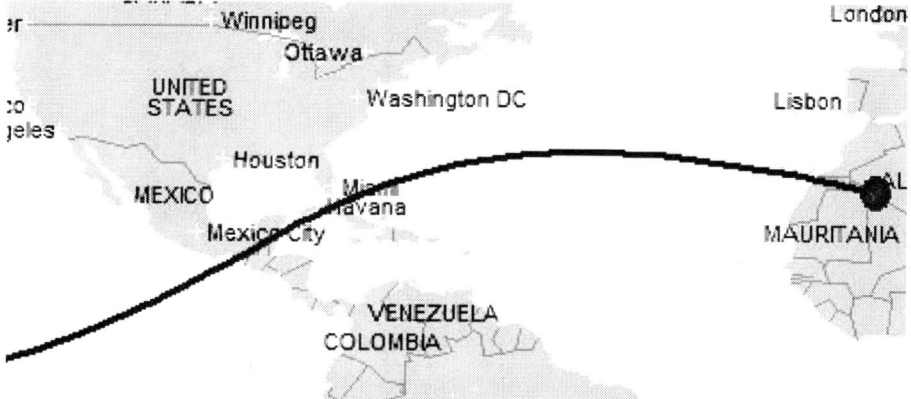

US-Mexican War The day after this eclipse passed over its southernmost
Solar Eclipse of territories Mexico was at war with the United States.
April 25, 1846 It started with a Mexican cavalry attack on a US patrol
in the disputed territories just north of the Rio Grande and was to end with the USA appropriating most of its present day southwestern states in the 1848 Treaty of Guadalupe Hidalgo.

These lands included California, Nevada, Utah and significant portions of Arizona, New Mexico, Colorado and Wyoming, some 500,000 square miles. Additionally the border of Texas was confirmed to the US advantage.

Although it saw the USA become a continental nation and a future Pacific power it was a poisoned chalice in that it led to a serious dispute between the non slave and slave states that was to end in the American Civil War just over a decade later.

The war, which Mexico was unlikely to win given that the eclipse line went over its territory, saw US forces march into Mexico City with little opposition. The partition of Mexico, indicated by its 1821 independence eclipse, had occurred.

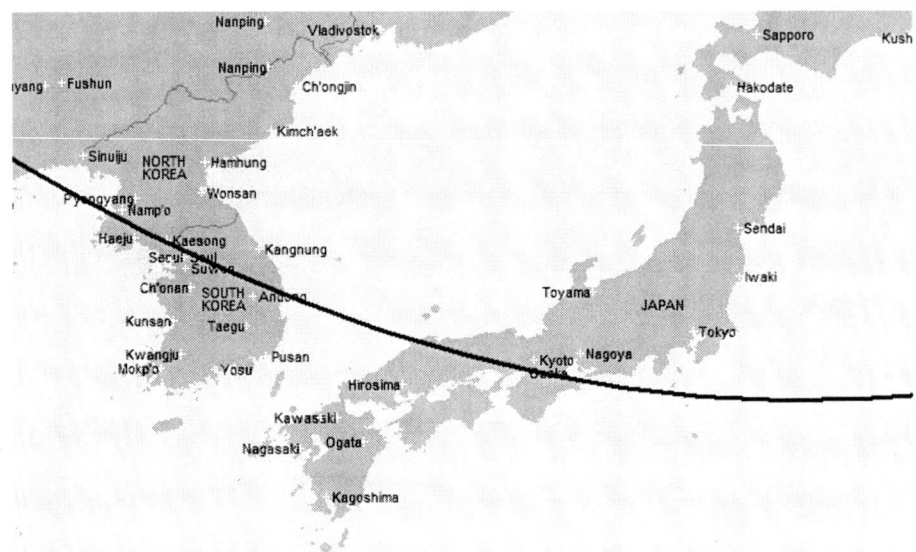

Commander Perry opens up Japan
Solar Eclipse of December 11, 1852

Just as this eclipse passed over Japan directly above the Emperor's palace at Kyoto, Commander Perry, of the US Navy, was steaming towards it intent on obtaining trading rights with the hitherto isolated nation.

His demands were met in full as the Japanese appreciated his "black ships" packed more firepower and were technically superior to anything they could assemble against him.

On March 31, 1854, 15 months after the eclipse, the Treaty of Kanagawa was signed guaranteeing protection to American merchants in Japan.

Whilst the USA demonstrated its naval strength it was a wake-up call for the "sleeping giant" Japanese who had kept themselves to themselves for the previous 200 years.

Smarting from their humiliation, they were to set out on a course of rapid modernisation leading to empire building and war with the USA nearly a century later.

No blood was spilled at the time of the eclipse but plenty was to be lost later as Japan established its place in the globalised economic, military and political world.

US-Canadian Border Eclipse of May 26, 1854

This eclipse line is interesting for the way that, given their Great Circle nature, it conforms very closely to the US-Canadian border.

No particular event occurred around it, coming as it did after the Oregon settlement eight years earlier but it is almost as if the heavens were confirming that there would be no more serious antagonism between the USA and Britain and Canada over their respective North American lands.

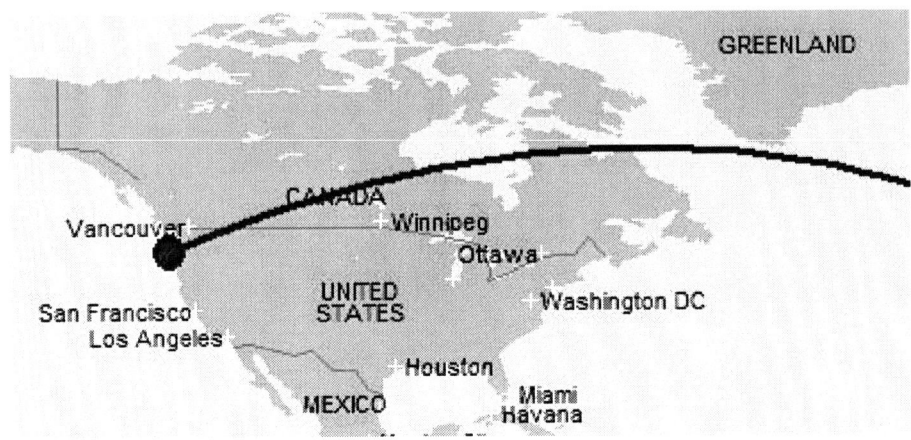

American Civil War
Solar Eclipse of July 26, 1860

This eclipse occurred less than 9 months before the start of the American Civil War between the Northern and the Southern States of the USA.

It is an encircling eclipse barely touching the United States at its line of totality although it would have been seen on a partial basis on what would become battleground sites.

Lasting four years the civil war ripped the heart out of the country and left a million casualties.

Ostensibly fought for the highest principle of ending slavery it led to much bitter carnage and a scorched earth policy in its later stages.

The industrial wage economy of the North was triumphant, not only establishing its dominance in the USA, but also enabling it to become the pervasive system worldwide.

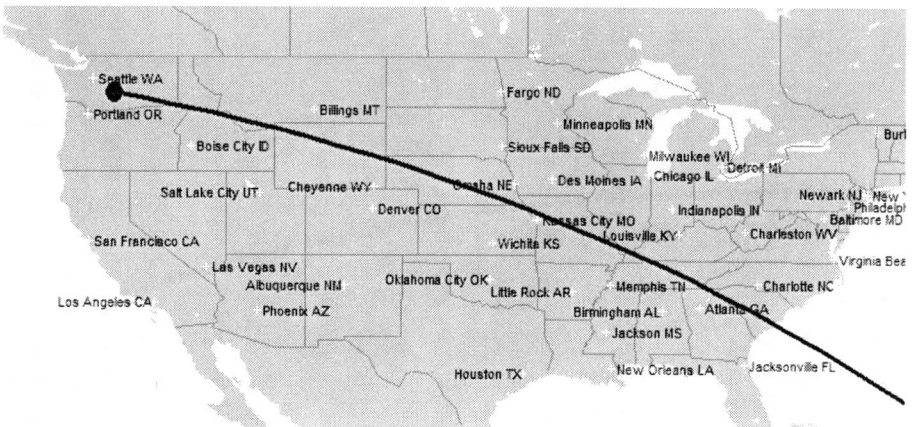

US-Western Indian Wars Solar Eclipse of October 19, 1865

This eclipse closely matches the line of the Oregon Trail, a 2,000 mile wheeled-wagon emigrant road across the plains from the Missouri Valley to Oregon.

Running North West to South East it dissected the USA a few months after the end of the American Civil War and the assassination of President Abraham Lincoln.

It indicated that the massive tragedy of the civil war would be followed by further travails.

These included the period known as Reconstruction in the Southern United States and also the re-enslavement of the freed Black Slaves who were denied opportunities for education.

It also saw the unleashing of the energies required to populate the West against the opposition of the indigenous peoples, who were greatly diminished during this period as the principle of private property was established over communal grazing and hunting lands.

Aided by the building of the first transcontinental railways thousands of Americans, whether homesteaders, ranchers or miners, moved into the Great Plains calling upon the US soldiery to protect them from the native Indians.

A core part of the USA's Manifest Destiny was the confinement of Indian tribes to restricted areas to facilitate more Western use of the land.

Many skirmishes were fought with tribes such as the Sioux, Cheyenne, Comanche and the Apache including those of the massacres of the Little Big Horn and Wounded Knee. Buffalo were also effectively extinguished in the wake of the eclipse.

US-Spanish War
Solar Eclipse of July 29, 1897

Less than nine months after this eclipse over Cuba the USS Maine blew up in the harbour of its capital city, Havana, precipitating the war between Spain and the USA.

It was to lead to Cuban independence from Spain as well as the USA's acquisition of the former Spanish colonies of the Philippines, Guam and Puerto Rico.

This eclipse marked the USA as a power which could control, not just the land areas of the Americas, but also the surrounding oceans of the Pacific, the Atlantic and the Caribbean.

China's Tragic Journey

**From Feeble Giant to Global Superpower
- how solar eclipses marked China's weakness and
afflictions in the 19th and 20th centuries and assisted its
eventual rise to global superpower.**

China on Notice
Solar Eclipse of
August 28, 1802

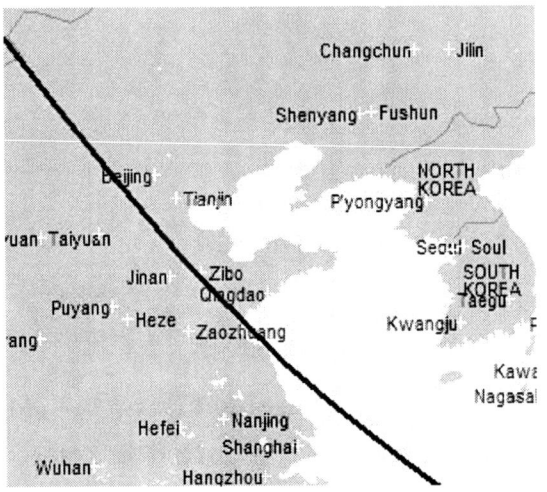

No dramatic event occurred around this eclipse but its passage over Beijing, China's capital city, and proximity to Shanghai, its greatest economic centre, indicated that the Middle Kingdom was no longer as powerful as it thought it was, as evidenced for the next two centuries.

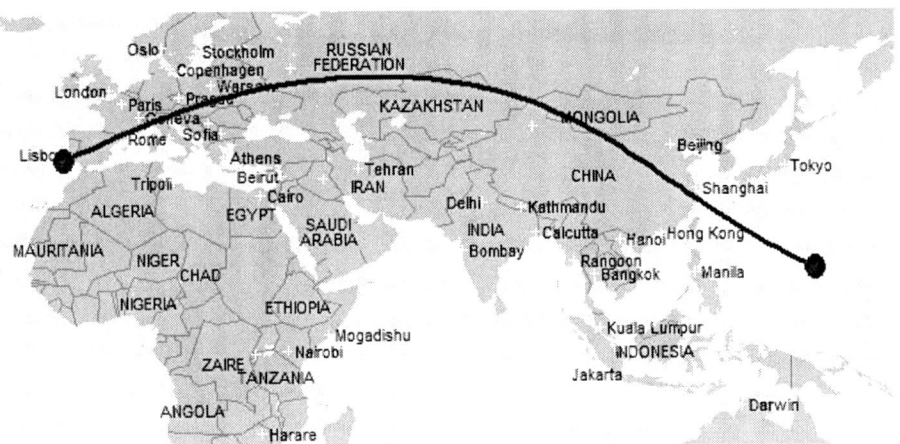

China opens
up to Europe
Solar Eclipse of
July 8, 1842

Just seven weeks after this eclipse linked Europe to China the Treaty of Nanking formally ratified the ceding of Hong Kong to the British.

The first of the "Unequal Treaties" following an Opium War, it was a stark significator of the weakness of China vis-a-vis Europe, a situation that was to last for the next 150 years.

From now on China would have to find its place in the globalised world created by the technologically superior Europeans.

Interestingly it started over the Portuguese capital of Lisbon, perhaps indicating the future eclipse of Macau by Hong Kong.

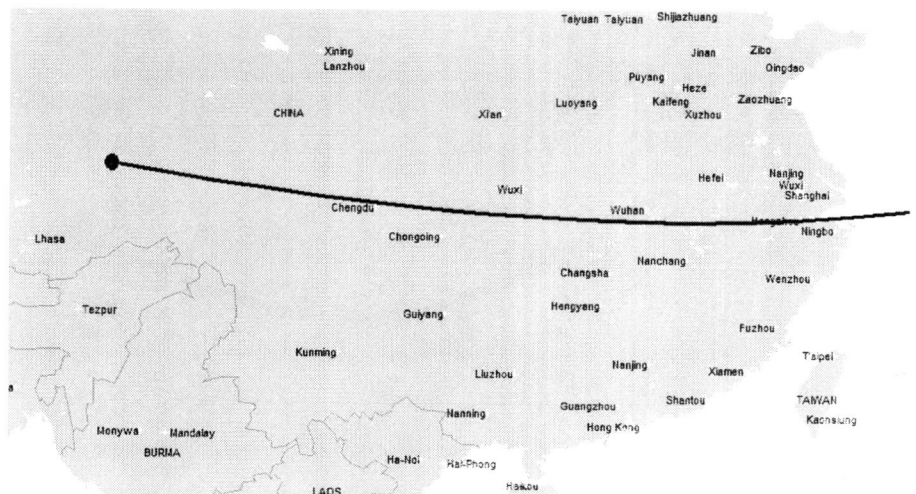

Taiping Rebellion
Solar Eclipse of February 23, 1849

Just over a year after this eclipse the initial shots in the Taiping Civil War were fired, leading to rebellion in 16 Chinese provinces and the deaths of an estimated 20 to 40 million combatants and civilians.

Some 600 cities were besieged and destroyed during the conflict which lasted until 1864 and mainly affected Southern China below Nanking.

Exposing the weakness of China's Qing central government, it was a total war with the destruction of agriculture a part of the strategies employed. Many of the fatalities, the largest number of any war in the nineteenth century, resulted from plague and famine. Over a million prisoners were massacred. Although Qing forces besieged the Taiping forces throughout the entire period of the conflict, it required the aid of British and French soldiery to finally suppress the rebels.

A feature of this revolt, found elsewhere in the aftermath of eclipses, was the Messianic derangement of its leader, Hong Xiuquan, who believed he was the younger brother of Jesus Christ and had himself pronounced the King of the Heavenly Kingdom of Great Peace (Taiping). He destroyed ancestral temples, banned alcohol, opium, prostitution and foot binding and proclaimed equality of the sexes and an equal distribution of land.

The 2007 film, The Warlords, featuring deception, massacre, misplaced loyalty, banditry, betrayal and Machiavellian intrigue, is based on this conflict, no doubt reflecting the real experiences of those caught up in it.

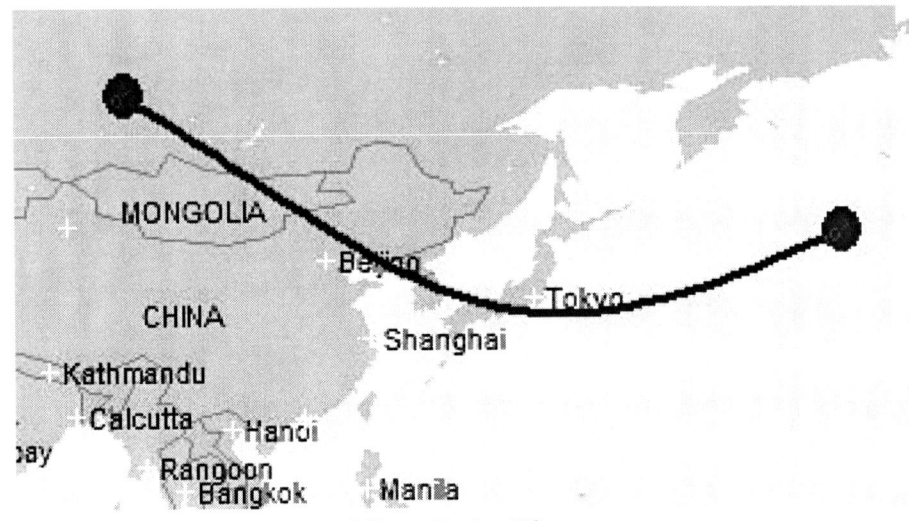

Nien Rebellion
Solar Eclipse of December 11, 1852

The Nien Rebellion, which began a year before this eclipse and lasted until 1868, was an uprising in Northern China parallel with the Taiping Rebellion.

It caused huge economic devastation and loss of life, despite having no clear objective apart from not being subject to the Emperor. Natural disasters and the inability of the government to deal with them added fuel to the flames of the insurrection as did the perception that the Qing Dynasty was craven in its relationship with the European powers.

Initially using cavalry attacks the Nien then fortified captured cities, resulting in constant fighting and the despoliation of vast tracts of formerly rich agricultural lands.

The Nien were finally defeated by Chinese forces armed with Western weapons.

Historians have often asked why the Nien and Taiping rebellions never linked up which would have made them far more formidable. A potential reason is that they were the result of separate eclipses.

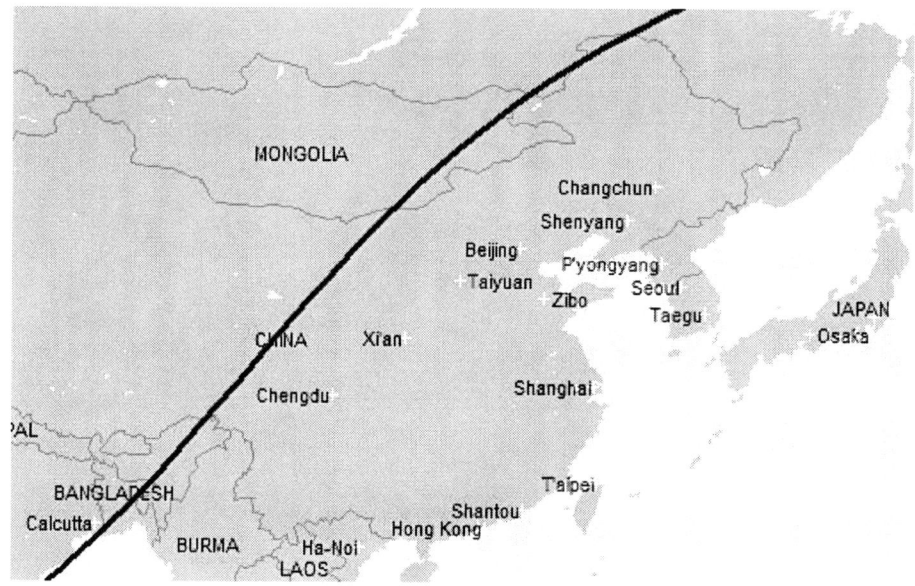

First Sino-Japanese War
Solar Eclipse of April 6, 1894

This eclipse swept over China four months before the ruling Qing dynasty's inadequacies were once again exposed in the First Sino-Japanese war. The diagonal bisecting of the eclipse indicated another downgrading was in store for the Giant of the East.

Whilst the Japanese had adopted best European military and naval practice with advice from British, French and Prussian counterparts the Chinese were still mired in corruption, inefficiencies and lack of application.

The conflict saw the Japanese trounce both the Chinese Army and Navy in separate engagements. The victories enabled Japan to take over, not just Korea, but also Taiwan. This signified a seismic change in the balance of power in the region which has only now been reversed nearly 120 years later.

The frailties of China's Imperial system led some to seek restoration of national pride through attacks on foreigners and others to insist the future lay with a Republic. Both approaches would be played out in the ensuing two decades.

The legacy from the conflict around this eclipse is still with us today with the division of Korea, the separation of Taiwan from China and the dispute between China and Japan over islands in the East China Seas.

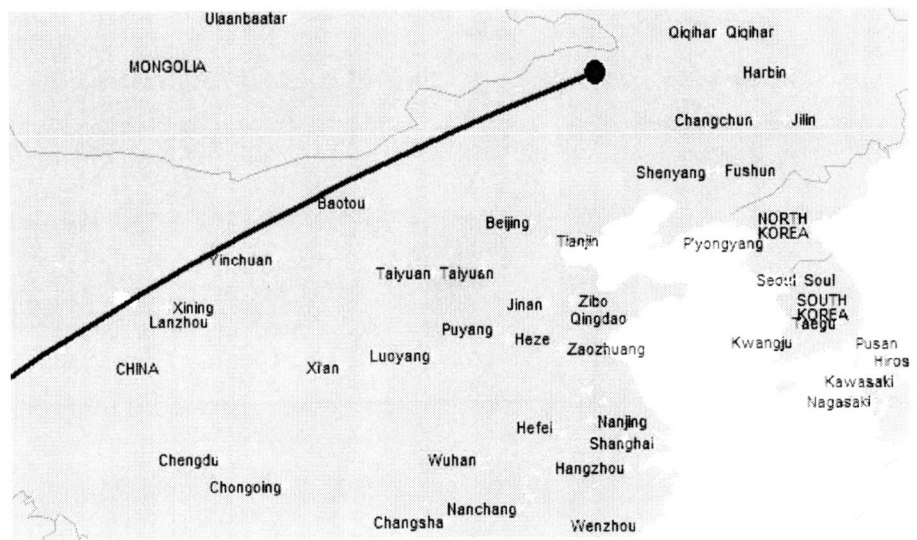

Boxer Rebellion
Solar Eclipse of January 22, 1898

Shortly after this eclipse the Righteous Harmony Society began agitating in North East China against foreign imperialism and Christianity. It was to lead to the outright insurgency known as the Boxer Rebellion in June, 1900.

Foreigners and Chinese Christians were placed under siege in the Legation Quarter of Beijing for 55 days prompting eight foreign national armies, including those from Britain, Japan and Russia to march on the Chinese capital.

An estimated 150,000 died in the struggle which saw China's weakness and divisions exposed once again and even greater foreign control of the once sovereign nation.

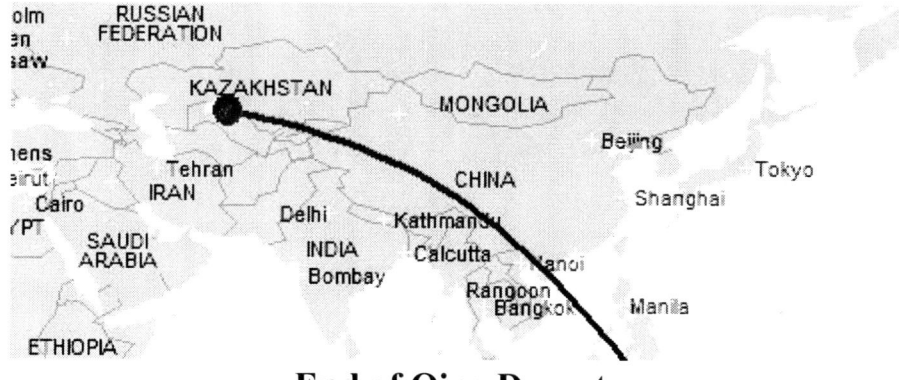

End of Qing Dynasty
Solar Eclipse of October 22, 1911

Coterminous with this eclipse the Qing dynasty fell and the Republic of China was declared. It was sparked by a small army revolt in Wu-ch'ang in Hupeh just 12 days earlier that won general support throughout the country and ended 2,000 years of Imperial rule.

It was to plunge the Middle Kingdom into even more turmoil and bloodshed, described as "brutal internal anarchy".

Chinese warlords, nationalists and communists vied for position to form a central government but the disunity enabled the Japanese, British and Russians to help themselves to significant portions of the country including Manchuria, Tibet and Mongolia respectively..

Japan usurps Manchuria
Solar Eclipse of April 18, 1931

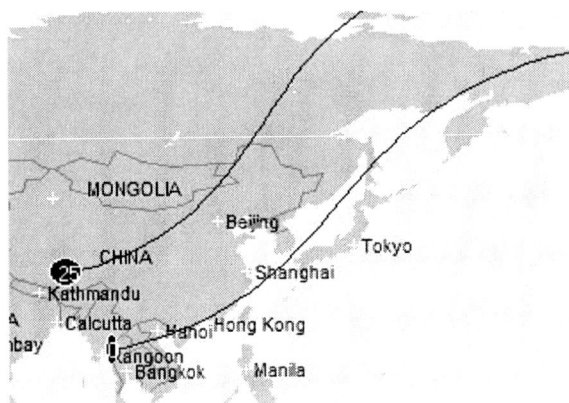

This partial eclipse preceded Japan's takeover of Inner Manchuria by five months.

It was not initiated by the Japanese government but rather by its army which staged the 'false flag' Mukden Incident to justify the territorial acquisition. The plot, which entailed the unsuccessful blowing up of a railway line, enabled the Japanese to protect its Korean colony from growing Chinese unity.

Outrage at Japan's duplicity and aggression led to its withdrawal from the League of Nations shortly afterwards.

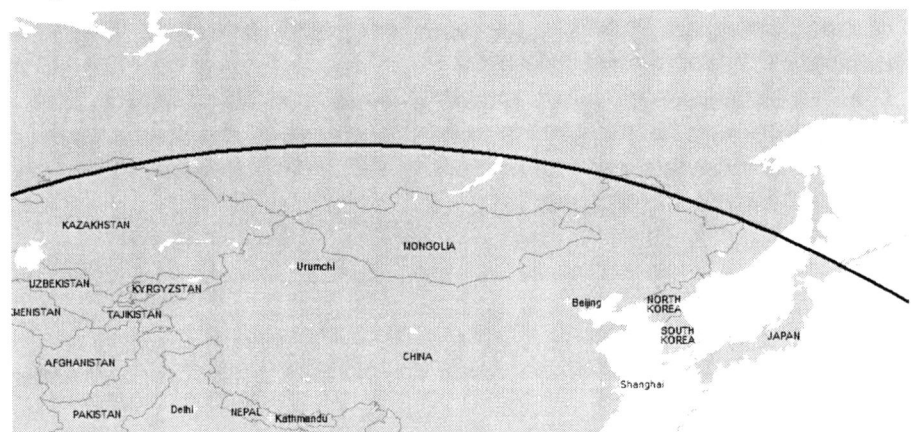

China Invaded
Solar Eclipse of June 19, 1936

Around 20 million lives were to be lost after Japan launched a full scale attack on China a year after this eclipse encircled both countries from the North.

The conflict, which continued after the defeat of the Japanese, between China's Communists and Republicans, was not going to be resolved for a further 13 years.

The brutality of the fighting and treatment of prisoners, including the use of Chinese prisoners for sword practise and sexual gratification, resounds to the present day.

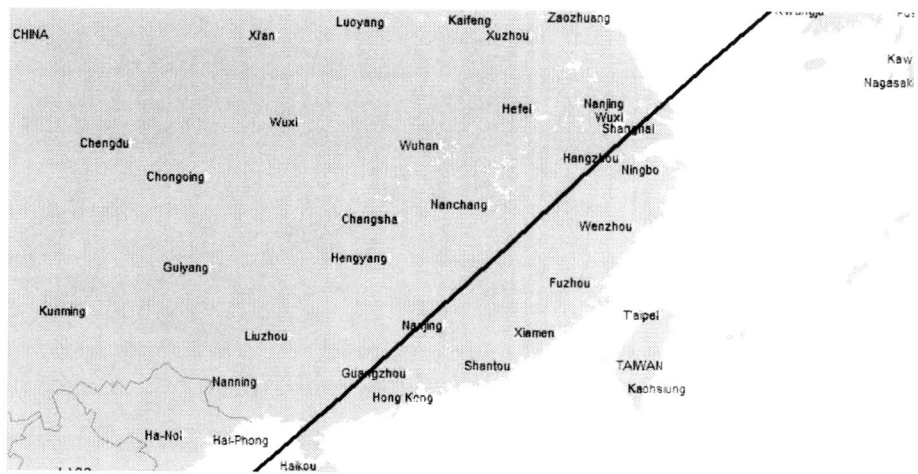

Fall of China's Nationalist Republic
Solar Eclipse of May 9, 1948

Just seventeen months after this eclipse traversed Eastern China the Republic fell and the communist People's Republic of China was declared.

It was a sea change in China's fortunes and in the world's. For the first time in a hundred years China had an undisputed centralised government capable of building it into a power commensurate with its population and land area.

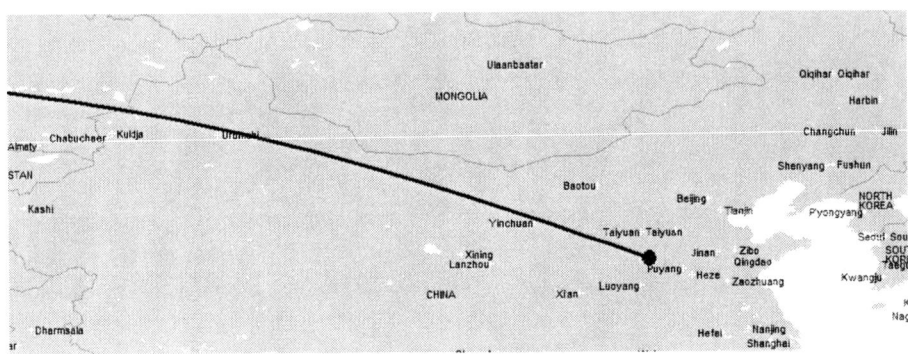

China's Cultural Revolution
Eclipse of May 20, 1966

A few day after this eclipse passed over China the "Cultural Revolution" took place witnessing the barbaric treatment of intellectuals and "class enemies" including humiliation, torture and murder.

Various figures for those who died range from 750,000 to 8 million, with an average of just under three million. As many as 100 million are estimated to have been directly harmed by the upheaval. China had become the callous plaything of Mao Tse-tung.

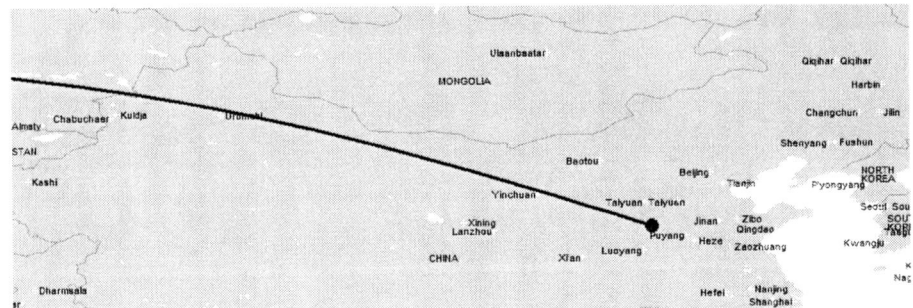

Death of Mao Tse-tung
Solar Eclipse of April 29, 1976

Five months after this eclipse stopped in West China Mao Tse Tung, "the Great Helmsman", was dead. Mao attracts the full range of opinion including that he was a merciless tyrant responsible for the deaths of 60 million of his countrymen and that he was the brilliant leader who unified his country.

Certainly China was placed in a strong grip similar to the Norman Conquest of England from which a powerful centralised state could make its mark on the world.

Chongqing - China's Modernisation Powerhouse Solar Eclipses of July 22, 2009 and January 15, 2010

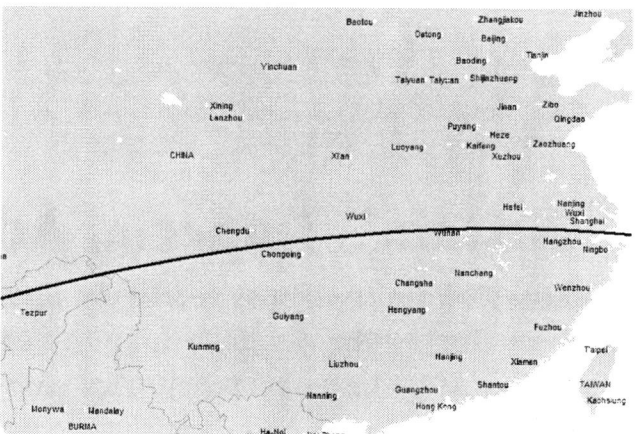

Both these eclipses went over Chongqing in West China, the world's largest construction site and fastest-growing metropolis.

It is possibly the world's biggest city with a population of 30 million and an economic growth rate in the lower teens per annum. Eclipses, whilst

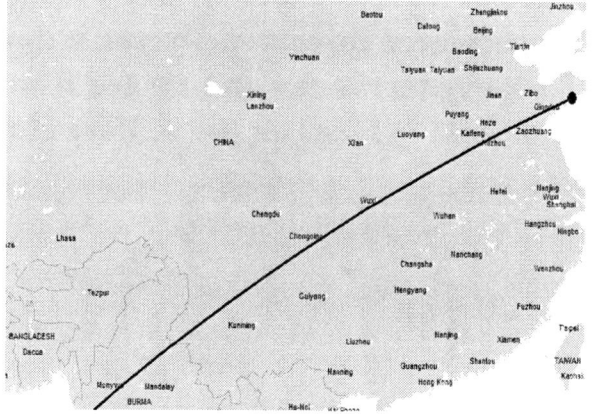

spurring intense activity, indicate that a venal legacy will result. China has moved to an economic war footing hell-bent on controlling the world's manufacturing, becoming more powerful but also more threatening to others.

The implications of China's rise for the West was evidenced recently by the scandal that rocked the Chinese establishment, A Harrow-educated British businessmen, Neil Heywood, was poisoned by the wife of the City boss of Chongqing, Bo Xilai.

Britain's influence on the world has been remarkable but now it seems to be coming to an unsavoury end, having met its match in terms of greed. Similarly the Chinese may well have swallowed an urbanisation fly.

China V USA Solar Eclipse of May 20, 2012
See Page 127

Britain's Imperial Adventure

Whilst Britain itself was relatively unscathed by eclipse activity during the nineteenth and twentieth centuries, a number of its imperial territories were overshadowed, thereby affording it an advantage to expand still further such as its acquisition of Hong Kong. The two eclipses it experienced directly delivered a significant break-up of its home territories and turned it from pursuing foreign conquest to focus on the welfare of its domestic citizens.

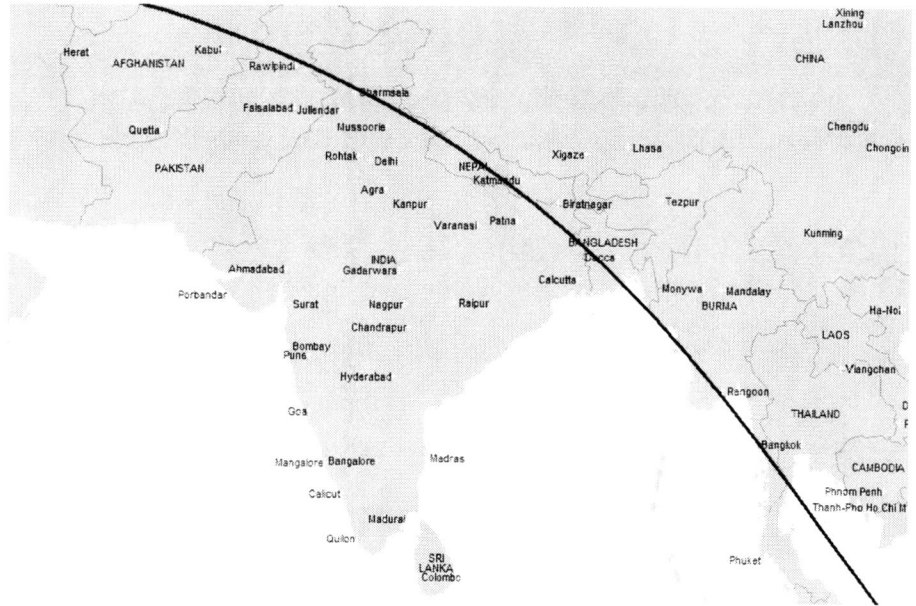

Indian Mutiny
Solar Eclipse of September 18, 1857

The Indian Mutiny started a few months before this eclipse swept through North India, the location of most of the fighting which included massacres of women and children, reprisal executions, the ending of the Mughal Dynasty and the dismissal of the East India Company as the ruler of India.

Normally eclipses occur ahead of the action but this one co-incided with an increase in severity. Just two days after this eclipse the British, in a move to stamp their authority on the region, executed the young heirs to the Mughal Empire. A scorched earth policy ensued.

Recent histories of the period emphasise the decade-long oppression in the aftermath of the Mutiny in which an estimated ten million Indians lost their lives.

Irish Home Rule

Eclipses marked the birth, violent uprising and achievement of Irish Independence after over 800 years of English rule.

Fenian Brotherhood Formed
Solar Eclipse of March 15, 1858

This eclipse over the UK appeared to have left it unmarked politically at the time, although it did foreshadow the publishing of the Origin of Species by evolutionist Charles Darwin, 18 months later.

However deeper research revealed that it co-incided with the establishment of the Fenian Brotherhood committed to an independent Ireland. The Fenians were to bedevil British politics through uprisings, demonstrations, guerilla warfare, assassinations, bombings and vitriolic dissension for a century and a half.

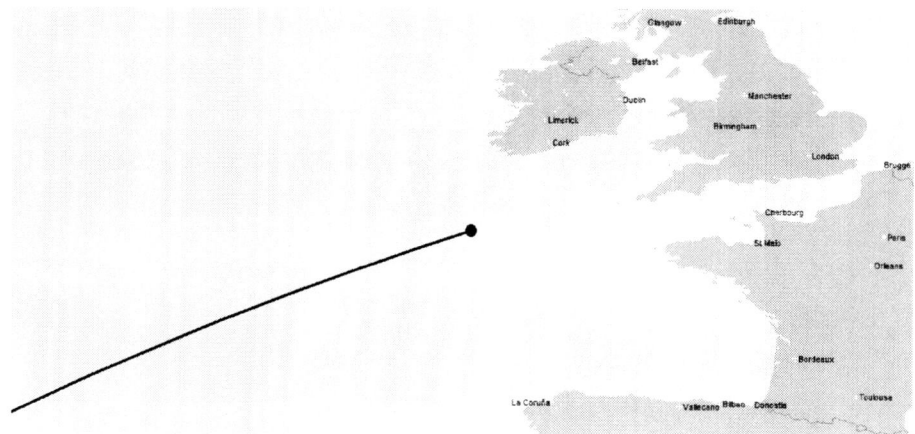

Easter Rising
Solar Eclipse of
February 13, 1916

Two months after this eclipse the armed rebellion of the Easter Rising took place, the culmination of decades of agitation for independence.

Eire Independence Achieved
Eclipse of April 8, 1921

Three months after this eclipse to the north of Ireland and Scotland a truce was established leading to Irish Home Rule.

Northern Ireland, however, remained loyal to the British Crown much to the annoyance of Republicans who continued to fight an insurgency to unify the island. The 1858 eclipse had achieved its diminution of the UK.

Southern Africa was the most colonised part of all of Africa by the British. The acquisition of its lands, the speeding up of its settlement, its expansion northwards and the 'Boer' civil war were all marked by eclipses over the southern part of the African continent.

Expansion into Queen Adelaide Province Partial Eclipse of July 7, 1834

Ten months after this partial solar eclipse over South Africa the British annexed Queen Adelaide Province now part of Cape Colony, around East London.

Its 7,000 square miles were the tribal lands of the Xhosa.

It was first time that the British attempted rule over an African people.

Diamond Rush Solar Eclipse of February 11, 1869

This eclipse co-incided with the onset of the "diamond rush" to South Africa which was to transform its fortunes from an agricultural to an industrial and urbanised economy. It was the dynamo of its economic fortunes.

By 1873 inland Kimberly, the centre of the mining industry, was the country's second largest town with a population of 40,000.

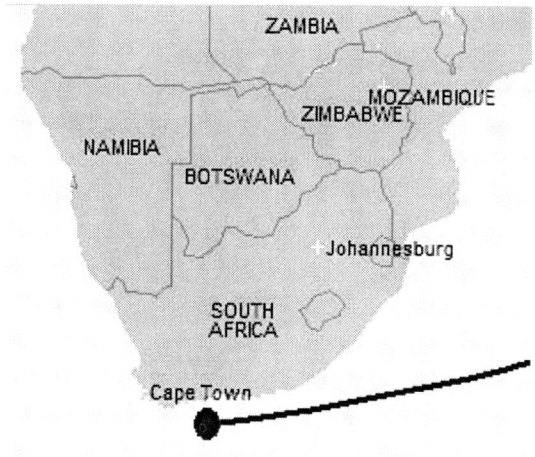

Annexation of Transvaal
Solar Eclipse of April 6, 1875

Two years after this eclipse the British annexed the Boer lands of Transvaal as much to protect the Dutch settlers from the native Zulus as to acquire new territories rich in precious stones.

The move was much resented by the Boers laying the foundation for their later conflicts with the British.

Zulu Wars
Solar Eclipse of January 22, 1879

This eclipse occurred at the exact time of the Anglo-Zulu battles of Isandlwanda and Rorke's Drift.

It was soon to be followed by the First Boer War between the British and the Transvaal Dutch farmers.

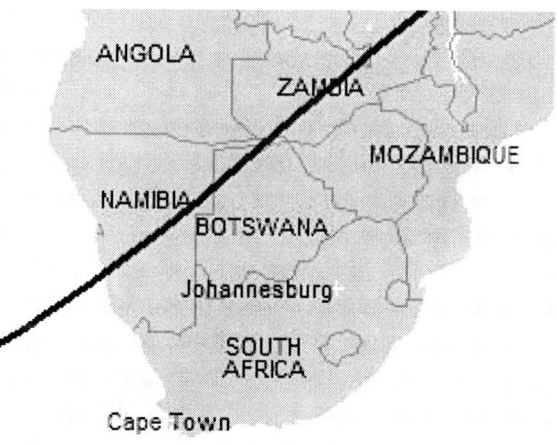

Move into Rhodesia
Solar Eclipse of June 28, 1889

A part of the Scramble for Africa series of solar eclipses, this eclipse, which is shown on page 94, saw the formation of Cecil Rhodes' expeditionary force and its movement north from South Africa.

The Column went on to annex the Rhodesias and Nyasaland for Britain, now Zambia, Zimbabwe and Malawi.

Boer War
Solar Eclipse of November 22, 1900

Although many of the main engagements of the Boer War had been completed by the time of this eclipse it occurred as it entered a new phase of inhumanity between the combatants.

A week after it Lord Kitchener became the British commander-in-chief in South Africa and introduced a more punishing approach to the Boer enemy.

It took the fight to the families of the Boer guerrillas with the intention of uprooting them.

A scorched earth policy was enacted which saw Boer homesteads torched, their wells poisoned, livestock slaughtered and their crops destroyed. Boer women and children were placed in concentration camps and surrendered combatants deported.

This was the first time that a concentration camp system had been used to target a whole nation leading to the depopulation of a large land area.

Of 28,000 Boer men captured as prisoners over 25,00 were transported overseas. Around 125,000 Boers were placed in the camps and over 25,000 women and children were to die in them of disease and malnutrition through overcrowding, poor sanitation and limited and unreliable food supplies. Many Africans experienced similar hardship in their segregated camps.

Some historians have suggested that the Boer concentration camps became the inspiration for the methods adopted by the German Nazis to carry out the Jewish Holocaust.

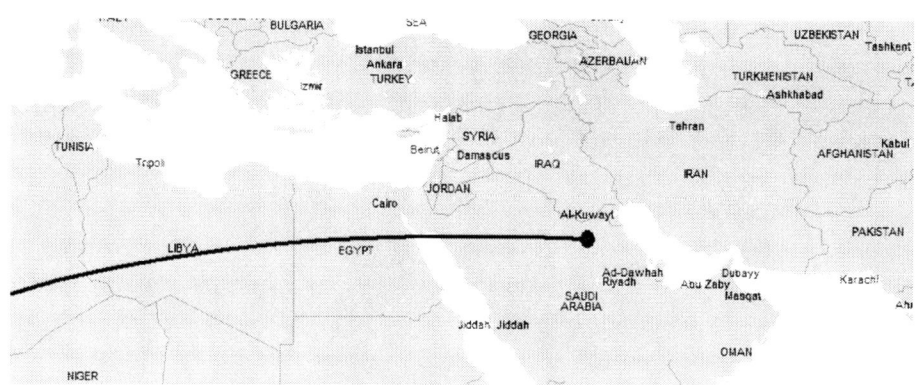

Palestine Mandate
Solar Eclipse of March 22, 1922

This eclipse co-incided with the tortuous settling of the political map of the Middle East following the defeat of the Ottoman Empire in which Britain played a major role. The so-called Palestine Mandate was a far from perfect solution and has left a stressful and aggravating legacy still ricocheting and causing upset in geo-politics to the present day.

Essentially the modern countries of Iraq, Jordan, Syria, Lebanon and Palestine (Israel) were established under a League of Nations Mandate which gave day-to-day control to the British and French. Additionally the rather vague desert borders of Saudi Arabia were made more definitive.

Britain, having had the benefit of strong island national statehood for centuries, was, once again, recreating the world in its own image under the belief that one type of political unit fits all.

In some respects this eclipse exposed the benefits of dynasticism as against nation statehood. In the one peoples of all races and religions could live together passively under the protection of a militarily strong family. In the other each ethnic or religious group became rivals seeking to benefit themselves in a competitive environment within a free-for-all democratic process.

All these relatively new states have suffered massive dislocation over the decades as the various peoples within them seek their selfish priorities. Syria is in the throes of a very nasty civil war; Iraq has been destabilised after Western invasion and state-building; Lebanon has been subverted by Iran; Israel is in a permanent civil war with the Palestinians it displaced and is constantly threatened with invasion whilst Jordan's benevolent Monarchy, is, almost on a daily basis, rocked by the crises of its neighbours.

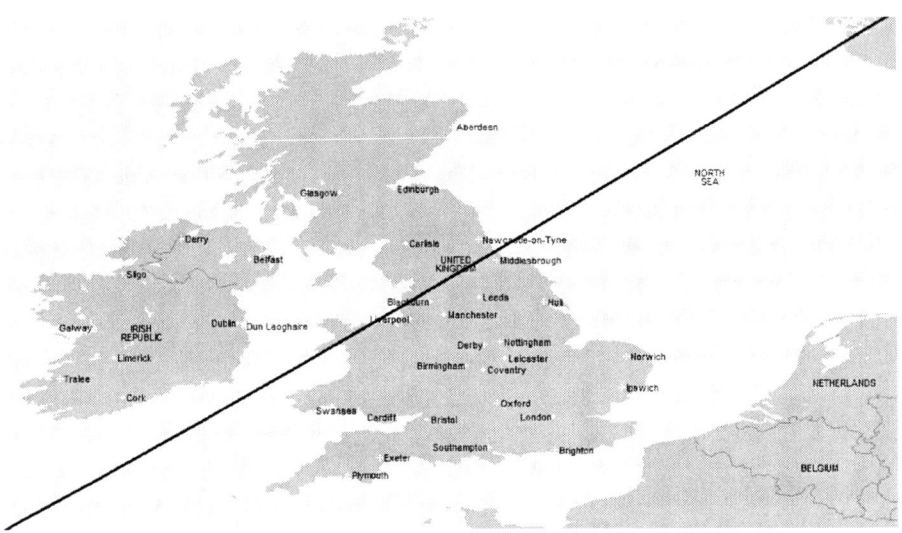

From Aristocratic Empire to Proletarian Welfare
Solar Eclipse of June 29, 1927

Whilst this eclipse passed peacefully over the United Kingdom it did mark a significant transition in its historical narrative. A few years earlier, with the Palestine Settlement, it had reached the pinnacle of the land area of its empire.

But now a new domestic power was rising which would make it more inward-looking. A year earlier the British trade union movement had bared its teeth with the General Strike and a few years afterwards Britain was to have its first Labour government and switch its direction to domestic welfare.

From being the workshop of the world and its chief coloniser it transitioned to focusing on the social well-being of its own citizens, a theme emphasised at the opening of the 2012 London Olympic Games with its portrayal of Britain's contemporary, more practical cathedral-building, the National Health Service.

The Falklands War
Solar eclipses of August 22, 1979, August 10, 1980
and February 4, 1981

Britain's continued capability to project sea power over long distances was exemplified by the April, 1982 conflict with Argentina over the Falkland Islands.

Three eclipses (not shown) over the islands and the southern south American country immediately preceded the conflict. Ranging from 25 to 75% in occlusion, Argentina was in greater darkness in two of them indicating its defeat.

Europe's 19th Century Dynasties and Nationalisms

Continental Europe in the 19th Century was first ravaged by the Napoleonic Wars and then by the forces of liberalism and nationalism which saw the rise of Prussian Germany, the constitutional chicanery of the Austrian Hapsburg dynasty and the fragmentation of the Ottoman Empire. On each occasion eclipses pinpointed the time and place of fundamental shifts in its political structure.

Fall of Holy Roman Empire
Solar Eclipse of February 11, 1804

Less than two years after this eclipse passed through the Eastern lands of the Hapsburgs the Battle of Austerlitz took place.

Won by the French under Napoleon against a joint Russian and Austrian force it led to a huge restructuring of the European political map including the dissolution of the 1,000-year-old Holy Roman Empire and the formation of the Hapsburgs' Austrian Empire.

Napoleon was so emboldened by his seemingly endless military victories that he crowned himself Emperor of the French shortly afterwards. This eclipse went over Russia, no doubt anticipating where his future ambitions would take him.

Greek War of Independence
Solar Eclipse of September 7, 1820

Six months after this eclipse Greek revolutionaries began the conflict that was to lead to national independence after a further 11 years of fighting.

Passing through Europe there were also popular uprisings in Spain which facilitated the independence of the Central and South American colonies.

This eclipse suggests that, not only was the Ottoman Empire's days' numbered, but the Monarchist Concert of Europe established by the Congress of Vienna five years earlier was going to find it difficult to maintain its grip.

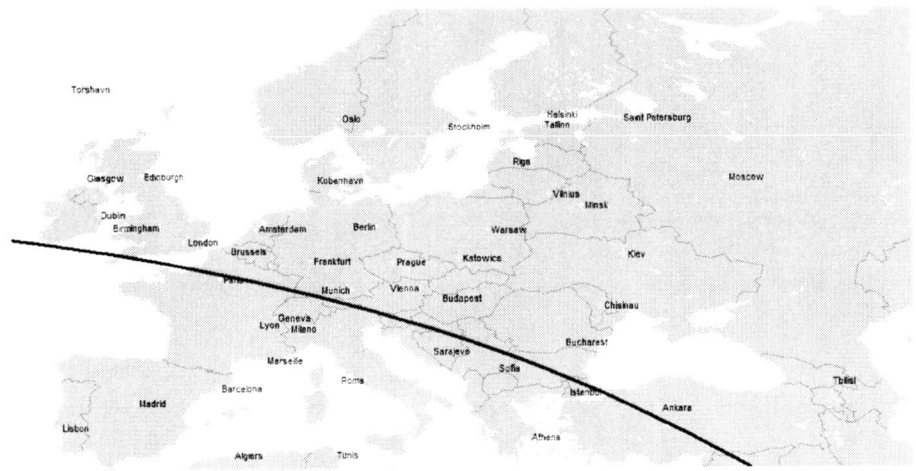

1848 Revolutionary Upheavals
Solar Eclipse of October 9, 1847

Four months after this eclipse passed directly over Paris the barricades were up again in its streets and the Orleans Monarchy of Louis-Philippe 1 was over. They went up again in June in bloody but unsuccessful revolution which was to lead to the establishment of the Second French Empire under Louis Napoleon.

Altogether rebellions took place in nearly every major European city during the year including Frankfurt, Berlin, Hamburg, Baden, Cologne, Vienna, Prague, Budapest, Venice, Florence, Milan, Rome, Palermo and Naples. However the forces of Conservatism proved well-placed to contain them.

The legacy, however, included the publishing of the Communist Manifesto by Karl Marx and Frederick Engels in the immediate aftermath after they made an analysis of the conflicting social and economic forces at work in the uprisings.

Creation of Dual Monarchy
Solar Eclipse of March 6, 1867

Twelve weeks after this eclipse swept through the Balkans just south of its territories, the Dual Monarchy of Austria-Hungary was inaugurated. A response to the Hapsburg defeat at Sadowa by the Prussians seven months earlier, it recognised that Austria was to be excluded from leadership of Germany and that it now had to concentrate on consolidating its patchwork empire of disparate peoples north of the eclipse line.

Passing through the principality of Serbia it also indicated that the Ottoman lands in the Balkans were going to become centres of new nationalist fighting. As if to impress this point Serbia itself gained de facto independence at this time as the Ottoman garrison pulled out of Belgrade.

Franco Prussian War and German Unification
Solar Eclipse of December 22, 1870

Within a few weeks of this eclipse over southern Europe a new power emerged in the north of the continent with the declaration of the German Empire, a merger of the Prussian-dominated Northern confederation with the southern states including Bavaria.

Occurring half way through the Franco-Prussian War the establishment of a unified Germany as a successor to the Holy Roman Empire was to dominate European politics for decades to come, even up to the present day.

After his defeat by Germany the Emperor Louis Napoleon of France was overthrown and the barricades went up again in the streets of Paris.

The Rise and Fall of Communism

Whilst capitalism arose organically from trade and entrepreneurism, especially in Europe, communism came from an intellectual and political re-action to its excesses and exploitations. An idea and nascent movement in the 19th century, it became a reality in the 20th, dividing the world, but subsequently found to be wanting. In a complete cycle, eclipses marked the birth of its founder, the events and writings that inspired its revolutionary nature, the births of its establishment in significant countries in Europe and Asia and the collapse of the Empire for which it had provided the ideological foundation.

Birth of Karl Marx
Solar Eclipse of May 5, 1818

Karl Marx, the founder of Communism, was born in Trier, Prussia, on the same day that an eclipse passed over Central and East Russia and encircled China, the two major countries to put his writings on Socialist Utopianism into practice.

In this respect he is similar, although not as long lasting, to Mohammad, who was also said to have been born shortly after the lands he was to have greatest influence upon were eclipsed.

Not only was his birth accompanied by an eclipse but so were his writings. Following the 1848 rebellions in Europe (see page 76) he and Frederick Engels published the Communist Manifesto, a call to arms for the burgeoning factory proletariat.

It was to become the inspiration for millions of revolutionaries around the world in succeeding decades and the justification for many deaths in the name of "Class War."

Six months after the 1867 eclipse over Europe (see page 77) Das Kapital, Karl Marx's analysis and criticism of capitalism, was published in German.

In it Marx states that Man lives in a world in which he is enslaved by the "monstrous power of inanimate money and commodities." It was to attempt to end this that Lenin orchestrated the seizure of power in Russia by the Bolsheviks fifty years later.

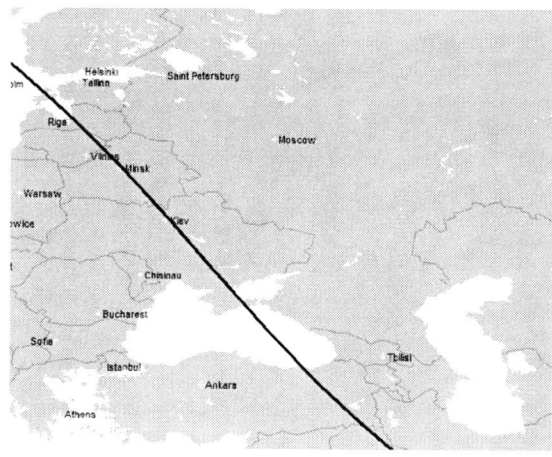

Tzarist Carnage
Solar Eclipse of
August 21, 1914

This solar eclipse across Western Russia, already discussed in the twentieth century wars section, sowed the conditions which precipitated the end of Tsarist Russia three years later paving the way for the Bolsheviks.

Fall of Tsar
Solar Eclipse of January 23, 1917

Two partial solar eclipses occurred over Russia in 1917 as Communist Bolshevism achieved power in two stages. First Russia moved from autocracy to Republic and then a few months later from Republic to Soviet.

The Tsar abdicated seven weeks after the first eclipse and the Menshevik government lost support after a failed attack on Austria in June at the same time as the second eclipse, enabling the pacifist Bolsheviks to put themselves at the head of the popular will.

Failure of Mensheviks Eclipse of June, 1917

It was to come at a great price though including civil war, foreign invasion and ceding of territory.

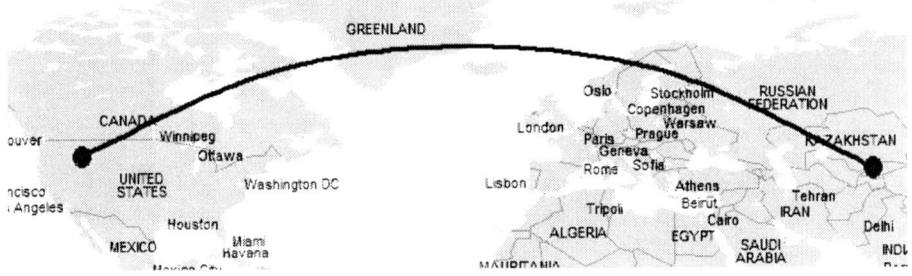

Cold War
Eclipse of July 9, 1945

As previously discussed Communist Russia and capitalist USA were the only game in town after VE day as this eclipse, which rose in the northwest of the USA and passed close to the Russian capital of Moscow, confirmed.

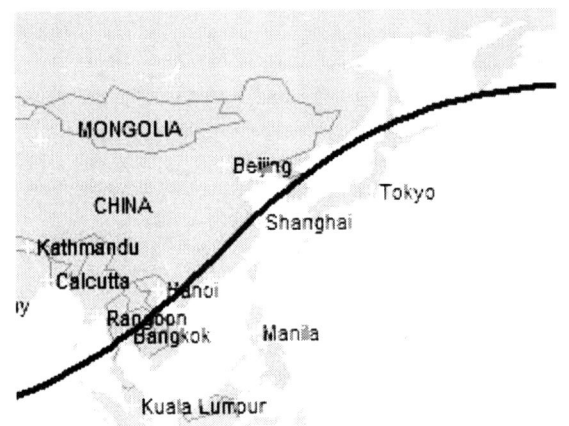

Pacific Asian Communism Solar Eclipse of May 9, 1948

This eclipse heralded the subsequent emergence of a number of new communist states on the East Asian mainland including China, North Korea and North Vietnam.

Collapse of Soviet Union
Solar Eclipse of July 22, 1990

Eight months after this eclipse encircled the Soviet Union, following almost exactly its northern boundaries, it began to disintegrate noticeably when Lithuania declared independence and seceded.

The death throes continued with a failed coup against President Gorbachev and secret deals between member states to establish a commonwealth of affiliates.

The Soviet Union formally ended 18 months after the eclipse, its unravelling leading to the break-up of Yugoslavia with its attendant wars and massacres as well as the creation of "frozen" states and wars in Chechen and Georgia.

Whilst communist rulers survived in less than a handful of cases in the Soviet region they remained in power in Asia including China, Vietnam and North Korea,

Most moved away from doctrinaire communism to state capitalism, although North Korea, still divided from the South, remained frozen in its Stalinist past.

South America - Independence and Nationalism

Early in the 19th Century the Spanish and Portuguese territories in South America sought independence. After its achievement they then had to establish their own national boundaries and forms of government. At each critical step solar eclipses marked their progress. And was a "strange" mass suicide in the north of the continent precipitated by an eclipse?

Rebellion and Fragmentation
Solar Eclipse of September 28, 1810

Eclipses marked both the beginning and the end of the wars of independence in South America generally held to have lasted from 1810 until 1825.

A further two occurred in 1813 and 1814 marking the crucial in-between stages of the Restoration and the renewed revolt.

Royalist Restoration
Solar Eclipse of July 27, 1813

Conflict Renewed
Solar Eclipse of
January 21, 1814

At the beginning of 1810 most of South America, apart from Brazil, was part of Spain's New World Empire. By 1825 Spain held only a few Caribbean islands. It was a fundamental shift in global power.

Initially the revolts began as

Independence Obtained
Eclipse of June 16, 1825

a response to Napoleon's conquest of Spain in 1808, thereby creating a power vacuum on the continent. Juntas were established in major cities that would become the centres of new countries such as Argentina, Venezuela, Ecuador, Chile, Peru and Paraguay.

With Spain at the centre it was possible to maintain the integrity of the continental possessions but local revolts led to the fragmentation of the territories.

This eclipse went through Quito which was a microcosm of the confusion of the early revolt. Initially leading citizens took power from the Royal representative on behalf of Spain's deposed King Ferdinand rather than submit to the new French Royalist rule.

This provoked a merciless backlash from the Royalist authorities which in turn engendered huge popular opposition enabling the Junta to approve a Constitution for Ecuador and embark on a campaign against the Royalists of Peru.

The 1813 eclipse co-incided with Joseph Bonaparte fleeing Spain and the Restoration of the Spanish Monarchy. Royalists loyal to King Ferdinand made renewed efforts to recover the rebellious territories.

It passed directly over the border of Peru and Chile and over Upper Peru (modern day Bolivia) where much of the fighting took place.

Similarly the 1814 eclipse, which followed a similar course across the centre of the continent, marked the realisation that a lengthy struggle would be required to attain the freedoms sought by the revolutionaries.

The 1825 eclipse started in Bolivia, just eight weeks before it formally obtained independence. This eclipse travelled mainly through Brazil which had a separate passage to independence than the former Spanish possessions.

Its war of independence ended ten weeks after the eclipse with the signing of the Treaty of Rio de Janeiro between the Kingdom of Portugal and the Empire of Brazil.

Paraguay's Self Destruction

Delusions are an integral part of life but the leadership of Paraguay made their country pay dearly in the 1860s for its belief that it could defeat its three much bigger neighbours in battle. An eclipse over a capital city is a very dangerous thing for its nation's citizens as is two eclipses in succession over the same area. Paraguay experienced both with drastic results for its population.

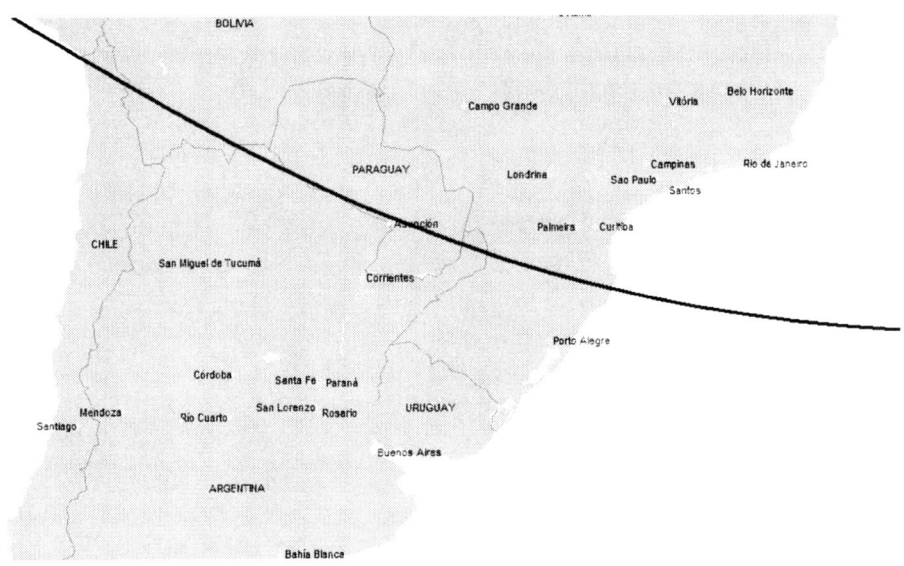

War of Triple Alliance
Solar Eclipses of October 30, 1864
and April 25, 1865

Less than two weeks after the first of these two eclipses went directly over its capital Asuncion, Paraguay had taken a Brazilian ship captive and prepared to declare war a month later on its much larger neighbour.

Not content with one David versus Goliath contest it then picked a fight with its other giant neighbour, Argentina.

One week after the second eclipse the three countries it went through, Argentina, Uruguay and Brazil, signed the agreement against Paraguay which

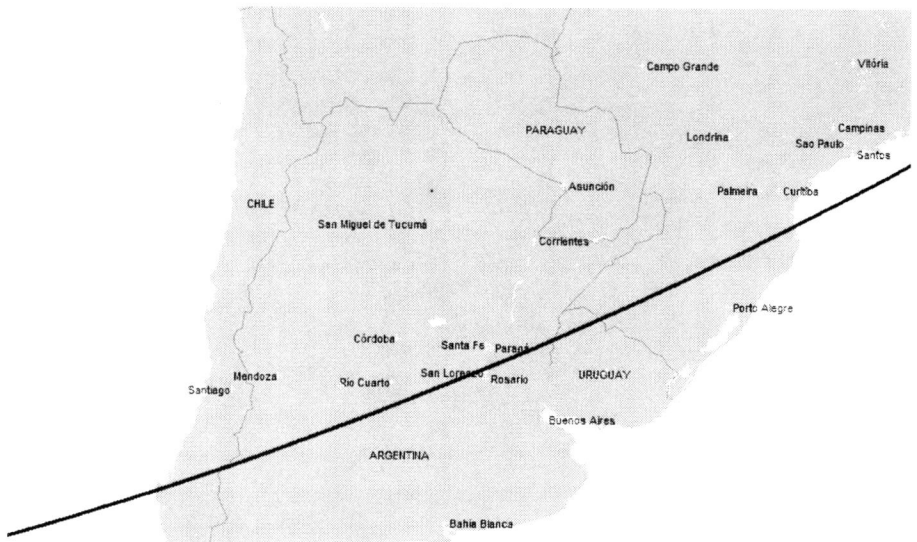

became known as the Secret Treaty of the Triple Alliance.

They were to deliver a terrible lesson to Paraguay which was to lose over 90% of its male population during the six years of conventional and guerilla war that followed.

Paraguay suffered more deaths per proportion of combatants than any other war in modern history. Additionally its economy was destroyed. Villages were abandoned and the survivors reduced to subsistence farming. Lands were sold to foreigners especially Argentinians who established large estates.

Previously self contained, Paraguay's markets were opened up to foreign trade and investment. Significant parts of its original territory were lopped off by the victors. They too suffered losses as the second eclipse suggested they would.

Brazil and Argentina got into serious debt to finance the war and Brazil had to abolish slavery soon afterwards. Additionally the Army, usually a bovine influence, became more involved in its national politics.

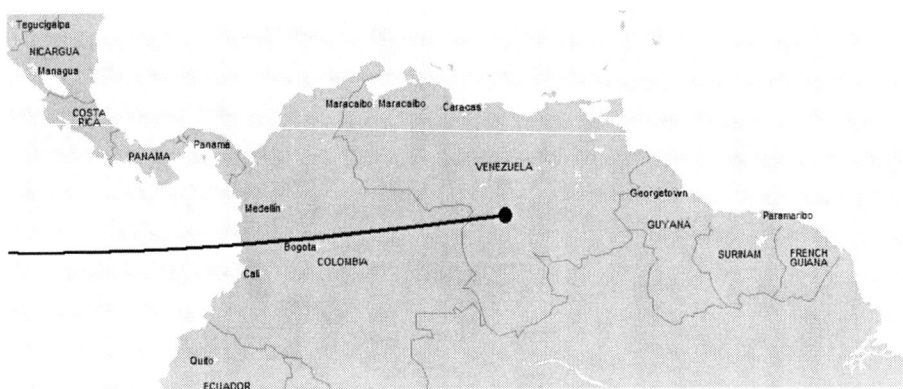

Jonestown Massacre
Solar Eclipse of
October 12, 1977

Thirteen months after this eclipse over northern South America, the world was shocked by a bizarre massacre of nearly 1,000 members of a religious group just a few hundred miles from where it ended.

The so-called 'Jonestown Massacre' occurred in a remote part of the Guyanan jungle where an American religious leader, Jim Jones, had established an Utopian community in which everybody worked together for the common good.

A visit from a US congressman to investigate conditions at the site led to shootings and a mass suicide in which members, including their children, drank a poisoned punch under instruction from their paranoid leader.

Africa's Colonisation and Decolonisation

Tribal Africa underwent two major transitions during the 18th and 19th centuries - takeover by European powers, followed by independence as new nation states. At each stage solar eclipses marked these rites of passage.

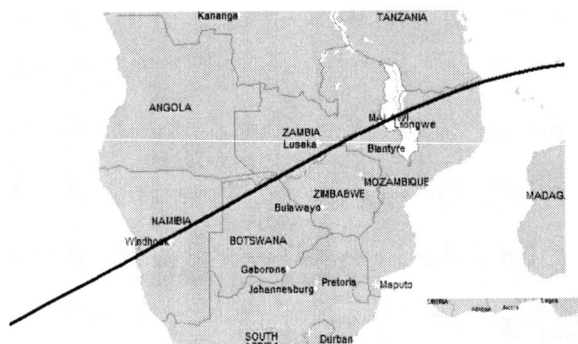

The Scramble
for Africa
Solar Eclipses of
June 28, 1889
December 22, 1889
and June 17, 1890

In just under a year, in a rare combination of three successive eclipses slicing through Africa, the European nations were likewise encouraged to portion out the continent amongst themselves.

Their "Scramble for Africa" had received triple shots from the heavenly starter's gun as they hastily and greedily sought to stake their claims ahead of their competitors.

First off in 1990 were the British led by Cecil Rhodes from their base in South Africa who moved into what became North and South Rhodesia and Nyasaland, later Zambia, Zimbabwe and Malawi. It was through these lands that the first eclipse had traversed.

Next off was the march of the Imperial British East Africa Company into the heart of Kenya and on into Uganda where the second of the eclipses had cast its shadow. At the same time the Germans took over Tanganyika and the British, the offshore island of Zanzibar.

Third off were the French who made a dash from the north into the Western Sahara, the path of the third of the eclipses. Moving towards it from the South the British Royal Niger Company pushed further inland into what became Nigeria.

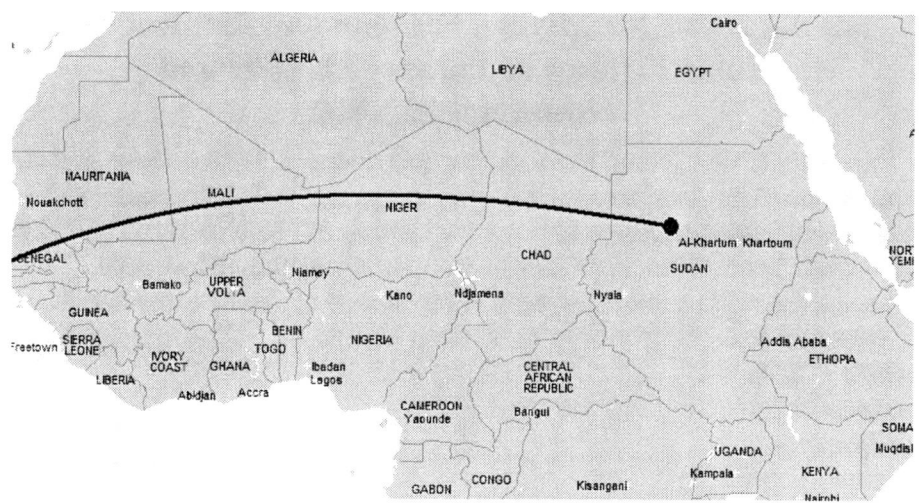

The Scramble for Africa Solar Eclipses of: April 16, 1893 and September 29, 1894

The April, 1893, eclipse passed over the Saharan region that France now felt able to assume as its colonies, marking a huge change in status for its resident peoples.

Following the formal annexation of the Ivory Coast a few months after it, French West Africa was established in 1895 and by the late 1890s, Senegal, French Sudan, Upper Volta, French Guinea, Ivory Coast and Dahomey were in the bag to be followed shortly afterwards by Mauritania and Niger.

The September, 1894, eclipse over Uganda, saw a massive change in the status of its peoples when it was absorbed into the British Empire as a Protectorate.

Designed to forestall a potential takeover by Germany, the area under British control was extended from the Buganda peoples to the North of Lake Victoria to the tribes and lands commensurate with the country today.

The Scramble for Africa - Fashoda and the Boer War Solar Eclipses of January 22, 1898 and November 22, 1900

Capping off the Scramble for Africa this January, 1898 eclipse saw the completion of the domination of the continent by the leading European powers. This is generally recognised as being the meeting of French and British forces at Fashoda in the Sudan eight months later just above this eclipse path.

France wanted to control Northern Africa from the Atlantic to the Red Sea in a West-East line whilst Britain intended to have uninterrupted territory from the North to the South, from Cairo to the Cape.

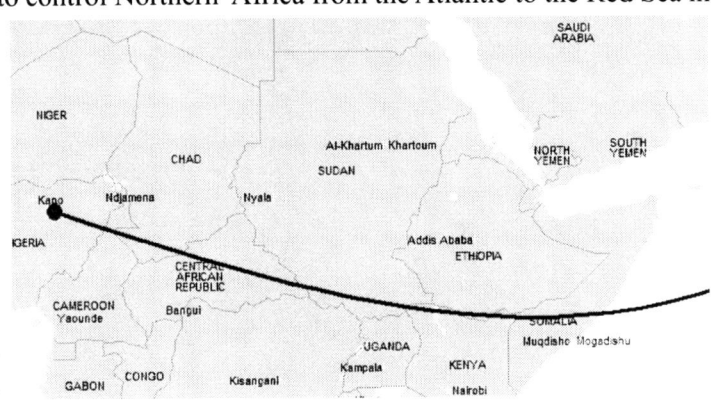

Fashoda meant that Britain would take the honours.

In many respects the European takeover was a type of incoming tide rolling over a flat beach meeting very little opposition.

In anticipation of the twentieth century wars the fiercest conflict of the overrunning of the continent came between contesting Europeans in South Africa in the form of the Boer War.

USA gains power in Africa
Solar Eclipse of September 1, 1951

The inevitability of African decolonisation was laid by this eclipse which connected the revolutionary nationalist, anti-colonialist heritage of the USA with Africa, then under European territorial control.

It ran from the USA's capital and power centre of Washington to cover much of the continent. Three months later Libya, formerly a colony of the defeated Italians, achieved independence.

Suez and Sudan Crisis of Empire Solar Eclipse of December 14, 1955

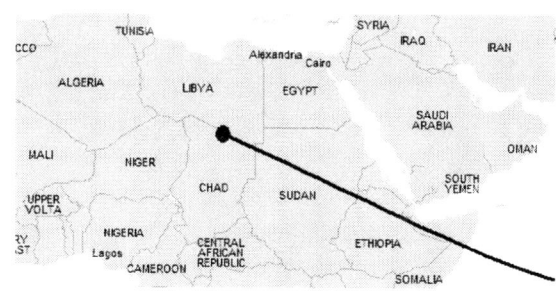

Within a few months of this eclipse Britain and France knew what the earlier 1951 eclipse had already ordained - that they could no longer call the shots in Africa, even with superior military force.

Although they were able to retake the Suez Canal, which had been nationalised by the Egyptians, by force, the power lay with the USA, which used its economic controls to force a climbdown by the embarrassed imperial powers.

This eclipse connected the independent countries of Libya and Ethiopia via the Sudan, which saw the start of its first civil war a few months earlier. It was to last for 17 years and is still reverberating at the present time, stimulated by later eclipses.

95

African Independence
Solar Eclipse of October 2, 1959

The 1944 independence eclipse for India and South East Asia had only touched Africa at its Horn. More time would be needed before the bulk of African colonies would be freed, 15 more years in fact for most of them when this eclipse occurred.

This eclipse was also connected to Washington indicating the influence of the USA. Following it a wave of independent African nations emerged as the French and the British, at the behest of the Americans, relinquished direct control.

Those countries obtaining independence along the line of and below the eclipse in 1960 included Somalia, Congo, Togo, Gabon, Central African Republic, Senegal, Cameroon, Upper Volta, Chad, Niger, Mauritania, Mali, Ivory Coast, Madagascar and Nigeria to be followed by Sierra Leone, South Africa and Tanganyika a year later; Uganda, Rwanda and Burundi in 1962 and Kenya and Zambia in 1963.

States in Southern Africa, however, well to the south of the main impact of the eclipse remained under colonial control longer.

Zambia and Malawi achieved independence in 1964 and Lesotho in 1966. Resistance against the inevitable came from Southern Rhodesia, which declared UDI and the Portuguese colonies of Angola and Mozambique which held out until 1974 after nationalist wars. A guerilla war led to the formation of Zimbabwe in 1980.

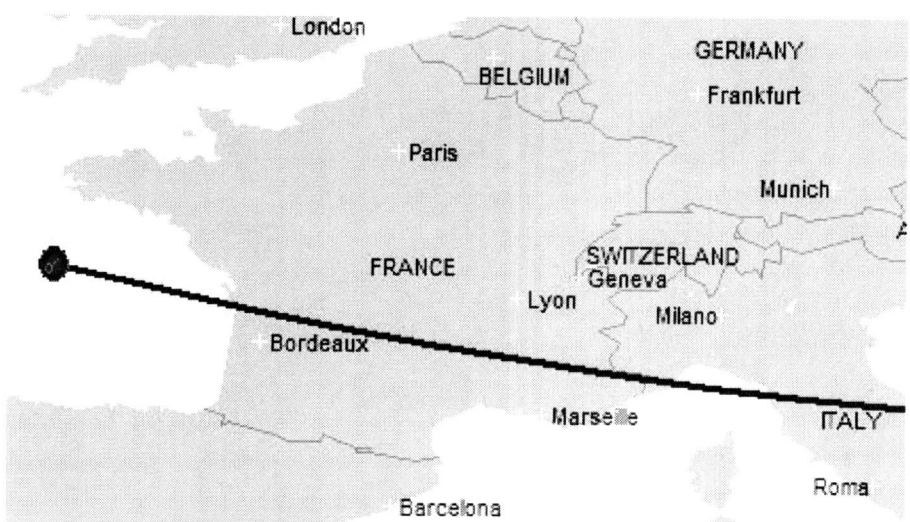

Algerian Independence from France
Eclipse of Feb 15, 1961

This eclipse over southern France occurred as President de Gaulle began secret negotiations to grant independence to Algeria, ending a conflict which had claimed nearly a million lives. It provoked a crisis in French society which had difficulties coping with further loss of prestige as a result of becoming a much-reduced Imperial power.

In less than a decade France had been forced to give up Indochina, been embarrassed by the Suez debacle and relinquished control of most of its West African colonies. Many Frenchmen felt Algeria was different. It was only the other side of the Mediterranean, had been a colony since 1830, had been settled extensively and was constitutionally part of Metropolitan France.

Only a massively patriotic figure such as de Gaulle was able to carry off the separation, narrowly avoiding a meltdown of the French national psyche.

The war of independence had been amongst the most vicious of its type with innumerable street bombings, massacres and assassinations causing many civilian casualties.

It included the "Paris Massacre" eight months after the eclipse when French police attacked a 30,000-strong demonstration by Algerians demanding independence. They beat some cornered demonstrators unconscious before throwing them into the Seine to drown. They also clubbed prisoners to death in the Paris police station courtyard.

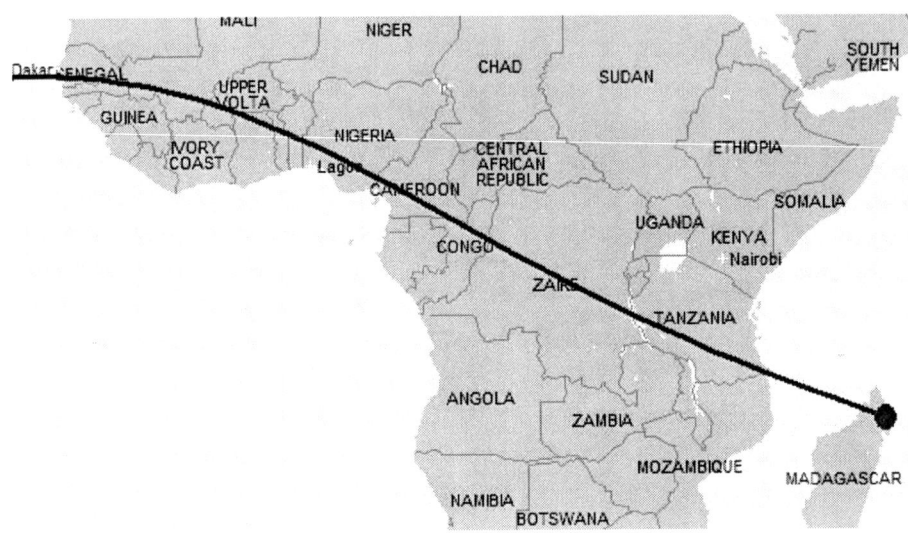

Africa's democracies diluted - Nigerian Civil War
Solar Eclipse of July 31, 1962

Two years after many African countries obtained their independence this eclipse punctured any false hopes that democracy was anything but a natural fit for the region.

In an indication that the principle of the separation of powers was going to be trampled over Senegal experienced its first serious crisis with the sacking of its prime minister and the acquisition of greater control by its president.

The first violent cracks in Nigeria's descent into civil war four years later occurred at this time. The making of a census, which would determine the numbers of MPs from different parts of the country, led to riots between the predominantly Islamic northerners and the majority Christian southerners as ludicrous, doctored numbers were announced.

Ominously this eclipse passed through the area of Nigeria known as Biafra which was to see starvation and genocide during the civil war.

Less than six months after this eclipse the first president of Togo was assassinated in a military coup d'etat.

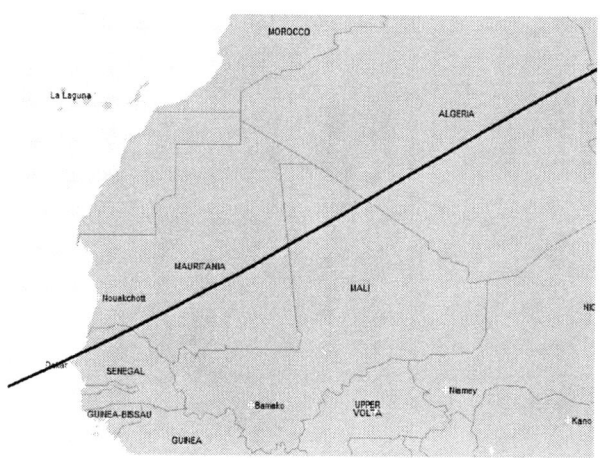

Polisario Insurgency Solar Eclipse of April 29, 1976

Just a few days after this eclipse over Saharan West Africa the Polisario movement began a large-scale guerilla war which was to inflict severe damage on the part of Spanish Sahara which had been taken over by Morocco.

Angolan Civil War
Solar Eclipse of April 18, 1977

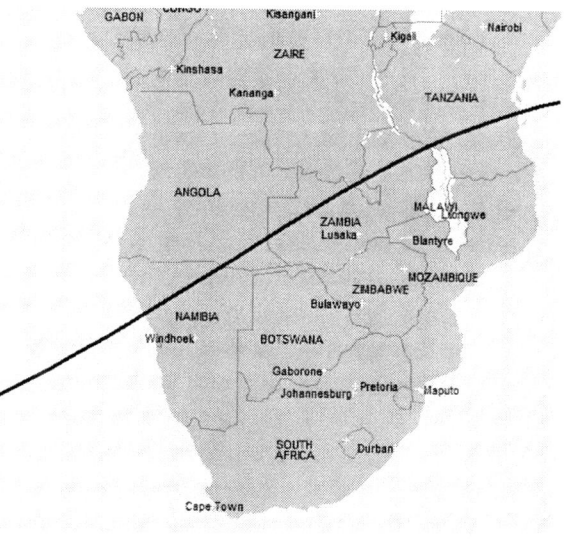

This eclipse occurred two years into the Angolan Civil War which was to last a further 25 years killing an estimated 500,000 people. It marked a significant escalation in intensity and cruelty in the conflict.

It passed over the Namibia-Angola border where Government and Cuban forces started to use flame throwers, bulldozers and planes with napalm to destroy villages and create an exclusion zone.

All males over ten years old were shot leaving only the women and children. The so-called Castro Corridor created a refugee crisis in neighbouring Zambia also under the eclipse path.

This eclipse also passed over the South Eastern corner of Zaire known as Shaba Province which had been invaded by an Angolan faction a few weeks earlier and was to be invaded again the following year.

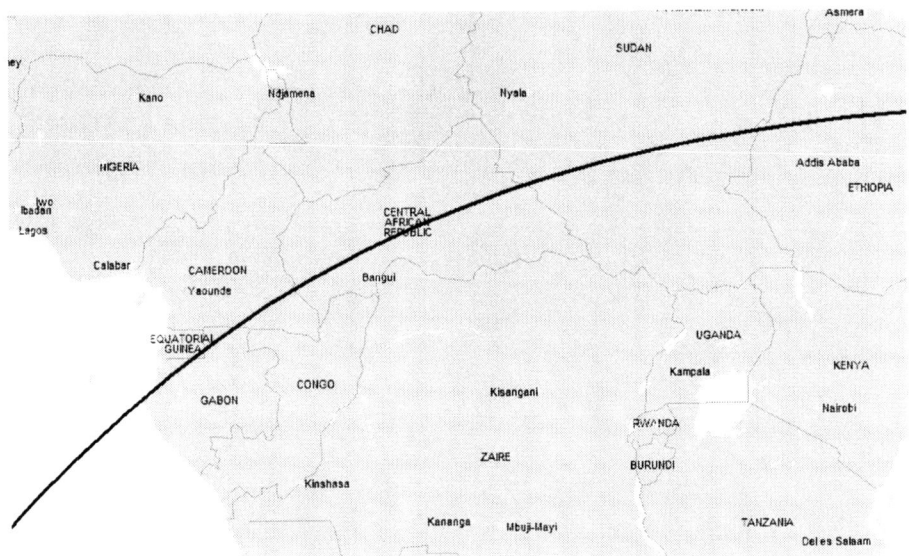

Lord's Resistance Army and Rwandan Genocide
Solar Eclipse of March 29, 1987

The Lord's Resistance Army sprung up in the immediate wake of this eclipse operating in the Central African Republic, Southern Sudan, the Democratic Republic of Congo and especially in Northern Uganda.

A religious movement it is a designated terrorist organisation, infamous for its child soldiers, estimated at numbering between 60,000 and 100,000. Amongst its violent acts are abductions, murders and mutilations. Its persecutions have lead to the displacement of two million people.

Its theocracy is an odd mix of Christianity, Islam, African witchcraft and tribal nationalism. It is led by one Joseph Kony who proclaims himself as the spokesman of God and a spirit medium.

This eclipse would also have created an 80-90 per cent occlusion over Rwanda, the location of a sudden eruption of genocidal massacres a few years later.

Case Study: The Sudan

As Africa's largest country, the Sudan has been particularly afflicted by wars and rebellions since the late 19th century, invariably timed by eclipses over its land area. In this case study of a single African nation recent solar eclipses indicate its partition with the creation of South Sudan is insufficient to end its perpetual strife and that it would benefit from being partitioned still further with an independent Darfur.

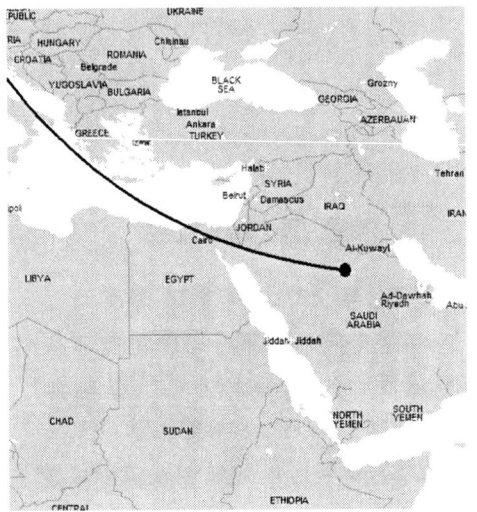

Egypt expands into Sudan Solar Eclipse of September 7, 1820

Just seven weeks before this eclipse over the capital of Cairo an invasion force was sent by the Egyptians into Sudan leading to its absorption a year later.

The "Turkiyah" was to be a bitter experience for the Sudanese who experienced greater taxes as a result of the occupation and an increase in slave-trading.

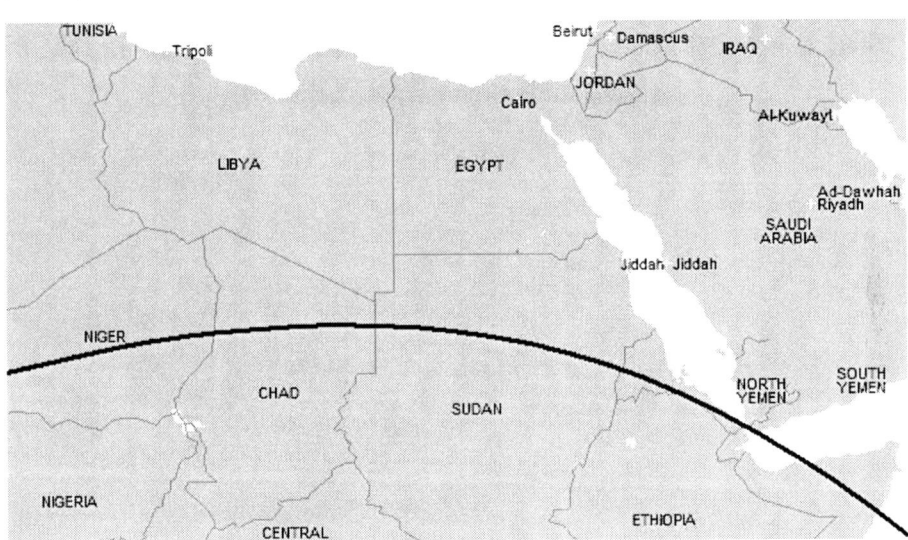

The Mahdi Revolt
Solar Eclipse of July 18, 1879

This eclipse, contemporaneous with the Orabi Revolt in Egypt, saw, within two years, the rise of an apocalyptic Islamic cleric, Muhammad Ahmed. Known as the Mahdi or "guided one", he waged Jihad or Holy War against the angrily resented Egyptian rule. Much of the Sudan was to fall under his and his successors' control for the next two decades.

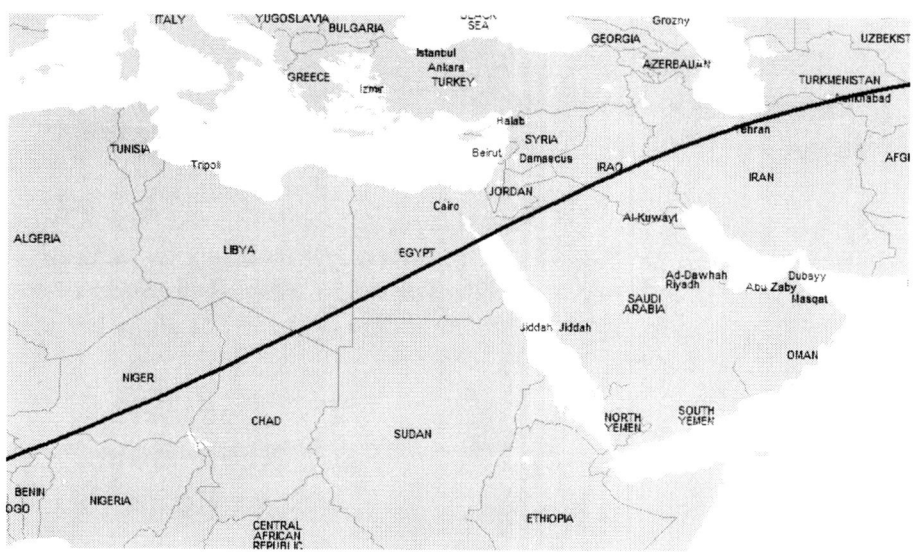

The Sudanese War
Solar Eclipse of May 17, 1882

At the time of this eclipse the Mahdist Army was overrunning the area around Khartoum prompting the ill-fated British Hicks expedition and the massacre of the garrison including General Charles Gordon. While the Mahdi was to die three years later his regime survived into the 1990s launching several Holy War assaults on its neighbours.

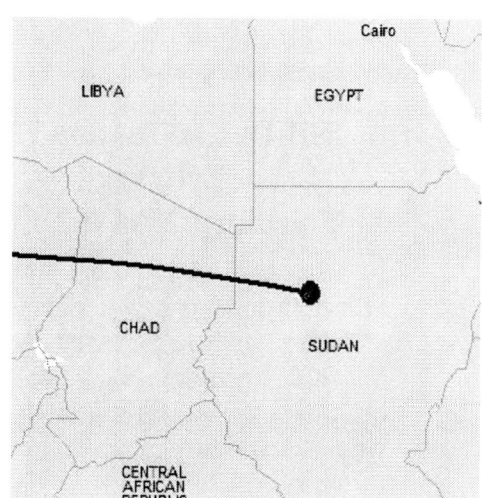

Decline of Khalifa
Solar Eclipse of April 16, 1893

This eclipse marked the decline of the Khalifa. Although it had been successful against the Ethiopians it had been defeated by the British in Egypt and the Belgians to the South.

The writing was on the wall for the religious militants and the British began planning their return which they accomplished two years later.

Fashoda and Omdurman
Solar Eclipse of January 22, 1898

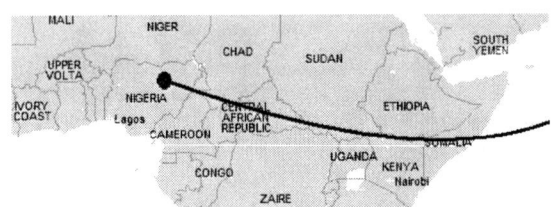

Widely seen as the end of the Scramble for Africa British and French forces backed off a conflict at Fashoda on the Nile a few months after this eclipse paving the way for the Entente Cordiale. The British established greater control under the Anglo-Egyptian Condominium following the defeat of the Mahdi at Omdurman in September, 1898.

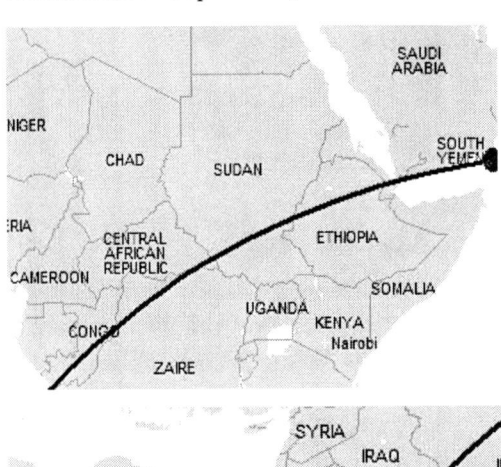

Sudanese Nationalism
Solar Eclipse of February 24, 1933

Whilst no dramatic events occurred around this eclipse in the Sudan, two years later neighbouring Ethiopia was invaded by Mussolini's Italy.

This was probably the time that Sudan first started to develop a desire for political independence and began to organise it.

Self-Determination
Solar Eclipse of February 25, 1952

The overthrow of the Faruk Monarchy in Egypt, a few months after this eclipse, preceded the Sudanese right to self determination as agreed in the Anglo-Egyptian accord of February, 1953.

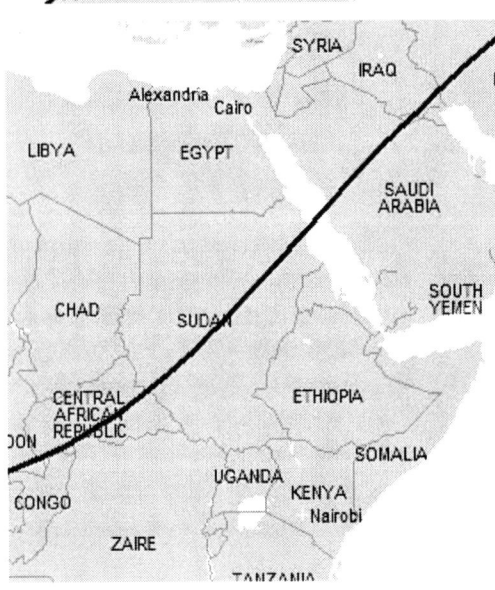

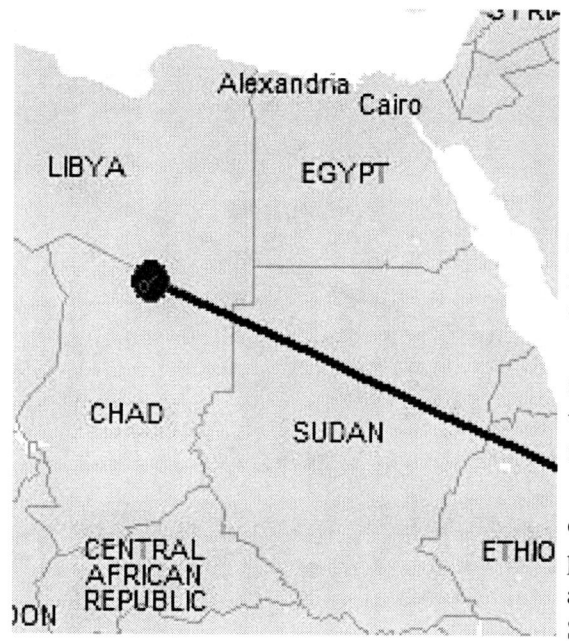

Independence and First Civil War Solar Eclipse of December 14, 1955

Two weeks after this eclipse Sudan gained independence from the Anglo-Egyptian Condominium.

The constitutional change sparked immediate conflict with the start of the First Sudanese Civil War.

This was to last 17 years with constant conflict between the peoples of the Islamic North and those of the Christian South.

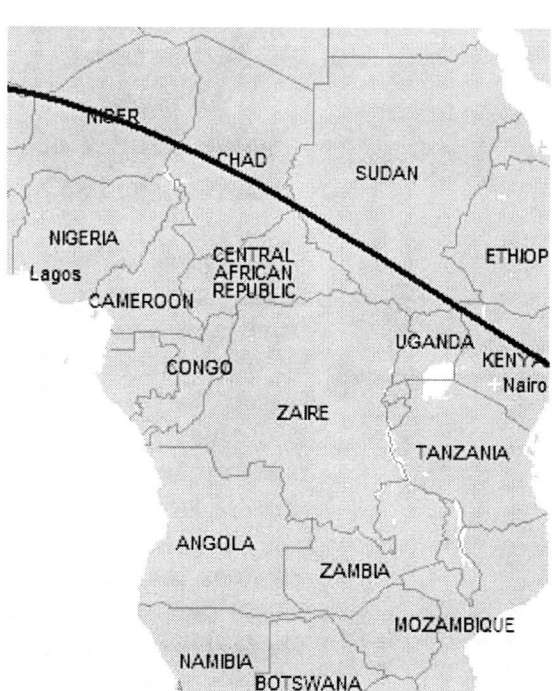

State of Emergency Solar Eclipse of June 30, 1973

Although the first civil war ended a year before this eclipse the seeds were being laid for the second.

At this time there were a series of strikes and coup attempts leading to the declaration of a State of Emergency and presidential rule a year after it.

A later coup attempt saw the killing of 700 rebels in Khartoum.

Second Civil War Solar Eclipse of December 4, 1983

Sudan's second civil war erupted around this eclipse, following the government's attempts to introduce Islamic Sharia laws alienating moderate Muslims and the Christian South.

Two and a half million people died in the conflict and four million were displaced before it ended 22 years later as the country experienced economic bankruptcy, inflation, war and famine.

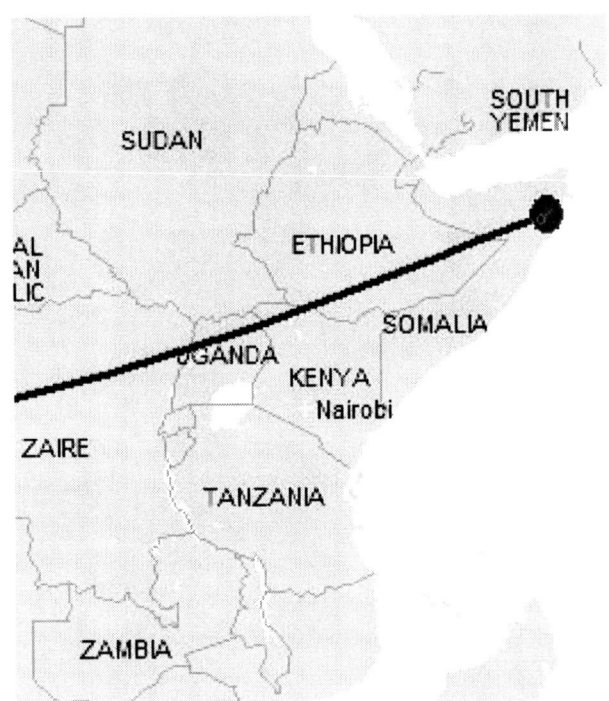

Sudan to be Divided Solar Eclipse of March 29, 1987

The future division of Sudan into North and South parts was heralded by this eclipse which bisected the country close to the subsequent boundary. Occurring in the midst of the ongoing civil war it anticipated the solution. At this time 20,000 "lost orphan boys" sought refuge in Ethiopia, most dying en route.

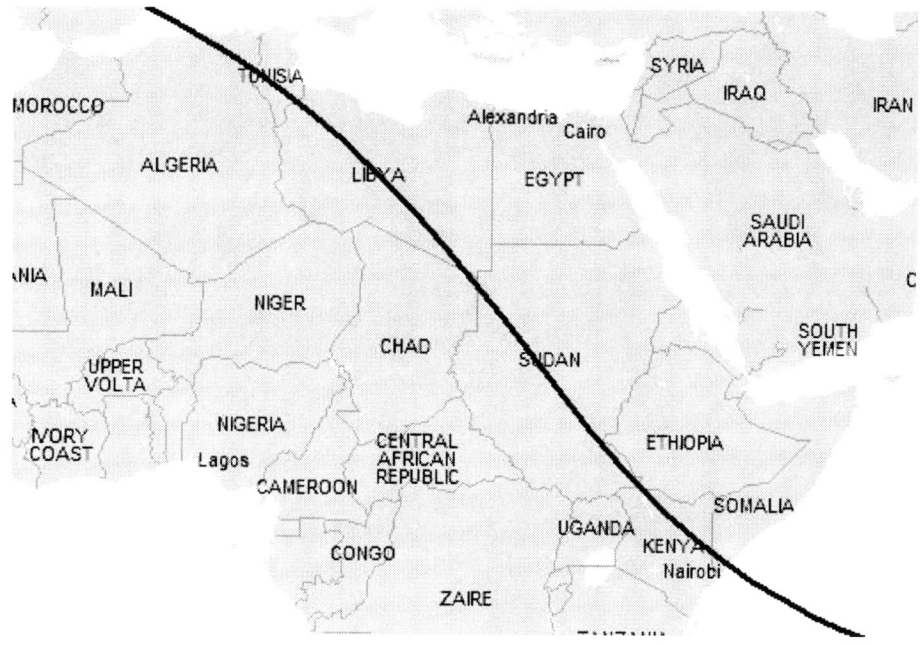

South Sudan Created
Solar Eclipse of October 3, 2005

Although the Second Civil War formally ended a few months earlier with a peace treaty granting independence to South Sudan, conflict in Sudan was not going to end so easily. Within 12 weeks of this eclipse Sudan was at war with its neighbour Chad.

This eclipse also cut through the area of Sudan where the so-called Darfur Genocide occurred. Racial rather than religious divides seem to be at its heart with the Arab-dominated government being accused of practising apartheid against their African co-religionists.

Although the conflict had started two years earlier it deteriorated rapidly at this time leading to the deaths of hundreds of thousands of civilians caught up in the conflict which saw mass starvation and widespread disease.

Mass displacements took place with many seeking shelter in refugee camps. At one time Darfur topped the United Nations list of those regions requiring food aid.

A ceasefire was agreed in 2010 but no agreement has been reached about a longer term solution.

Islam's Holy City Eclipsed

An 1803 eclipse went over Islam's holy city of Mecca bringing with it a significant, albeit brief, change to its control.

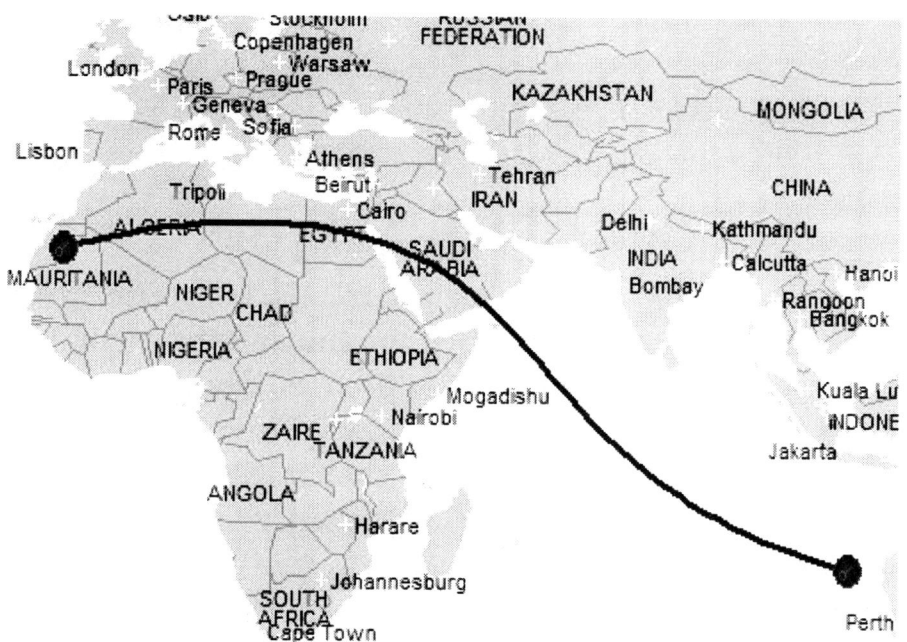

Mecca conquered by 1st Saudi State
Solar Eclipse of August 17, 1803

As the site of the birthplace of Mohammad and the revelation of the Koran, Mecca is the most holy city of Islam.

Whoever controls it has very high status within the Islamic community and also a very important responsibility.

At the beginning of 1803 it had been part of the territories governed by the Ottomans for nearly 300 years. At its end it was under the occupation of the First Saudi State which retained it for a decade, when the Khedive of Egypt retook it on behalf of the Ottomans.

The Saudis regained Mecca a second time in 1924, two years after the 1922 Palestine Mandate eclipse. It remains under their control.

Boundary Eclipses

Some solar eclipses have been particularly precise, given the 93 million miles between the sun and the earth, in their pinpointing of boundaries between countries; others in their division and secession.

These include eclipses, already shown, which went over Mexico in 1821, the US-Canadian border in 1854, Eastern Europe in 1914, Korea in 1948, the two that went over Vietnam's Demilitarised Zone in the 1950s and those that went over the Sudan in the 1980s and 2000s. Two others reflect current boundary disputes.

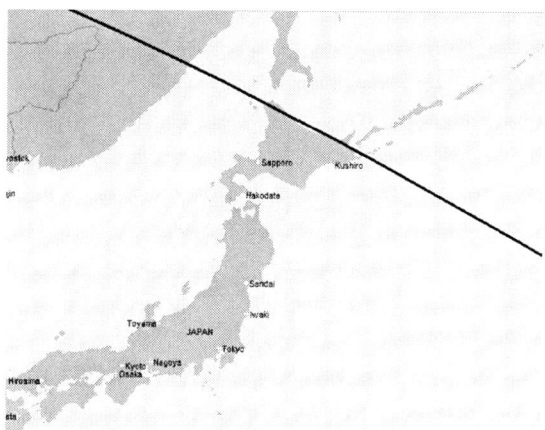

Sakhalin and Kiril Islands cut off Solar Eclipse of June 19, 1936

The Second World War eclipse bisected Japan's northern islands which would subsequently be annexed by Stalin and remain subject to dispute with Russia to this day.

Abkhazia cut off from Georgia
Solar Eclipse of March 29, 2006

Two years after this eclipse line divided Abkhazia (shown as Suhumi) from Georgia it was literally cut off from the Caucasus nation by Russian military action.

With South Obsettia it is now a frozen state recognised by Russia and a handful of small allies.

Eclipses and Dynasties

Prior to the Enlightenment and the promotion of the concept of the benevolent despot, a number of European Kings claimed to rule by Divine Right. God gave them the authority to control, exploit and punish their subjects. Charles I of England, Wales, Scotland and Ireland provoked civil war and his own execution by holding such an attitude, but the "Sun King", Louis XIV, of France, enjoyed considerable prestige by his similar approach. Do the heavens play a role in the fortunes of rulers? We have already seen how eclipses marked the end of the Mughal, Qing, Hapsburg, Romanov, Hohenzollern, Osman and Qajar dynasties in India, China, Europe and the Middle East. They also played an observable role in the rises and falls of France's Bonapartes.

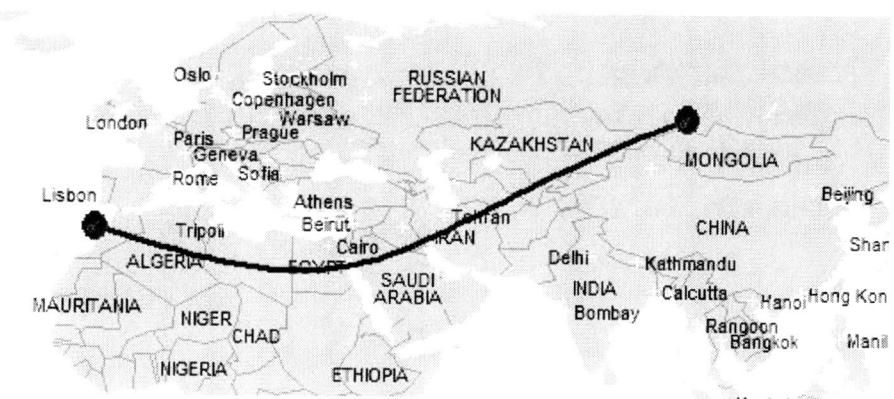

Fall and Death of Napoleon Bonaparte
Solar Eclipses of February 13, 1813
and August 27, 1821

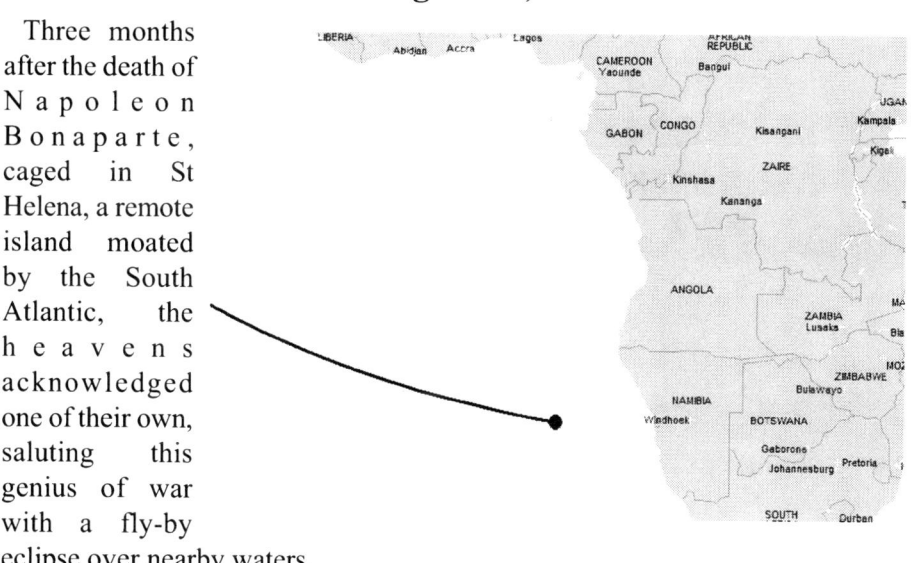

Three months after the death of Napoleon Bonaparte, caged in St Helena, a remote island moated by the South Atlantic, the heavens acknowledged one of their own, saluting this genius of war with a fly-by eclipse over nearby waters.

The self-proclaimed Emperor of the French had won more battles than Alexander the Great, Julius Caesar and John Churchill combined, requiring the whole of Europe - Russia, Austria, Prussia and Britain - to unite against him before he could be overpowered.

We have already come across this eclipse as that of the Monroe Doctrine and the seeding of the Texan, US-Mexico and American Civil Wars. The fact that it

Rise and Fall of Louis Napoleon Solar Eclipses of October 9, 1847 and December 22, 1870

connected to the individual who had destroyed Europe's peace for over two decades was a passing of the martial baton from one continent to another.

Eclipses were to mark the lives of Napoleon's heirs in a similar way to his own fall from power. He became Emperor shortly after the European eclipse of 1804 which destroyed the 1,000-year-old Holy Roman Empire and fell for the first time in 1813 following the February eclipse the same year in nearby North Africa and his subsequent defeat at the Battle of Leipzig.

In like fashion his nephew Louis Napoleon III was to restore the Corsican family dynasty to the French throne in the wake of the turbulence of the 1848 revolutions following the 1847 eclipse over Paris and to fall in 1870 after the Franco-German war/ German unification eclipse.

Death of Louis Napoleon's Son Zululand Solar Eclipse of January 22, 1879

The dynasty was effectively ended when Louis's own son died fighting in the Anglo-Zulu war a few months after the 1879 Isandlwana eclipse over Zululand.

Eclipse Personalities

As mentioned in earlier sections the births of
Mohammad, Alexander the Great and Karl Marx and the
death of Mao Tse-tung occurred around eclipses. The
lives of other historical and living individuals have also
been connected to eclipses including the times of their
assumptions of power.

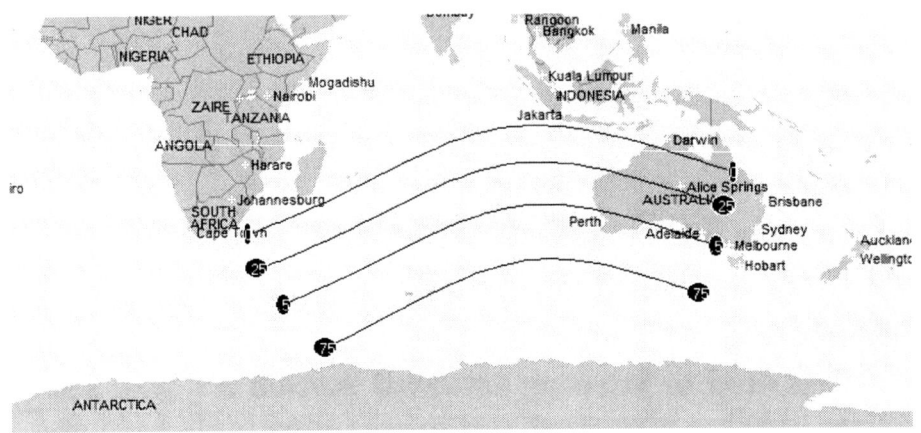

Birth of Pope John Paul 11
Eclipse of May 18, 1920

Three assassination attempts were made on the life of Pope John Paul II despite his advocacy of love and forgiveness. The failed assassins came from diverse groups including the communists, the Islamists and his own church indicating how indiscriminate he was in his capacity to make enemies.

Brave, great-souled and principled, even he couldn't prevent his eclipsed birth attracting people to him who wanted to kill him.

Many believe he was a true history-maker responsible for the freeing of Poland from the Soviet straightjacket, literally parting the Iron Curtain and allowing the capitalist and religious lights to shine in.

The godless system of Karl Marx, who considered religion to be the opium of the people, was finally laid to rest by the Catholic workers of Poland inspired by the Jesus-intoxicated Pope. It had taken an individual with an eclipsed birth to inspire the world-shaking cause of communism and a second individual with an eclipsed birth to bring it tumbling down. Both were born in May with their suns and moons in Taurus, the sign of the fixity of earth and associated with the Pope card in the tarot. Marx can easily be described as the pope of material socialism and Pope John Paul II, was, well, a pope.

His final succumbing to death was played out on world television as if it was a global sacrifice and an example on how to face mortality comparable to that of his church's founder. More heads of state attended his funeral, which also took place at the time of an eclipse, than any other person previously, no doubt in recognition of his contribution to reducing geo-political tensions and his genuine spiritual authority.

116

Death of
Billy the Kid
Solar Eclipse of
January 11, 1880

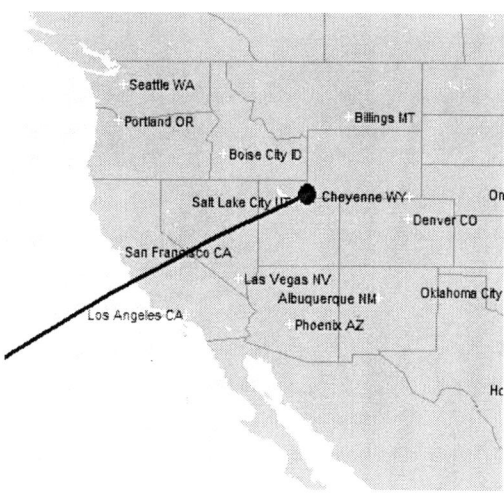

Oddly, some killers receive public sympathy and become folk legends.

Examples would have to include Robin Hood and Bonnie and Clyde.

One such anti-hero was William H. Bonney, a cowboy, rustler, gambler and outlaw who earned the nickname Billy the Kid after the governor of New Mexico put a price on his head in 1881 and newspapers took up his story.

Said to have killed between four and 21 people he was dispatched by a former friend turned sheriff, Pat Garrett, in July, 1881, a few months after this eclipse ended by darkening the "Wild West" of the USA.

Numerous books and films have been made about his life and death, which reflect the characteristics of an eclipse's ability to plunge previous innocents into a life of kill or be killed.

Another legend to appear at the place and time of this eclipse was Wyatt Earp, a lawman who famously fought a gunfight at the O.K. Corral in Tombstone, Arizona, in October, 1881, in which three "cowboys" were slain.

In both cases questions were asked about whether justice was being served by the deaths of those involved.

Death of Che Guevara
Eclipse of November 12, 1966

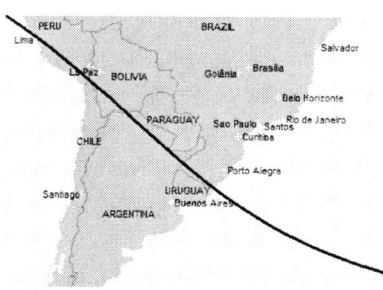

This eclipse passed over South West Bolivia a few miles from where the legendary Marxist guerilla Che Guevara was to die eleven months later attempting to spread revolution throughout South America. His execution was to lead to his idealisation as a martyr to the cause of left wing revolution and the ubiquity of a poster showing him as a hero.

117

Birth of Prince William
Eclipse of June 21, 1982

Prince William, the second in line to the British throne, was born on an eclipse day.

The premature tragic death of his mother, Princess Diana, known to have affected him deeply, was implied by his eclipsed birth.

The assumptions of power of Russian President Vladimir Putin at various times have been marked by very powerful era-making solar eclipses, reflecting his country's global alliances against the USA and the West.

Putin takes power
NATO-Islamist Solar Eclipse of August 11, 1999

Two days before this epoch-making eclipse which pitted the West against militant Islam, Vladimir Putin first came to public notice by being appointed the acting prime minister of Russia. Five days after it, he was confirmed as full time in the post.

He became acting president a few months later and has dominated Russian politics ever since, serving two terms as president, a further four years as prime minister before returning again as president for a third non-consecutive term in May, 2012.

The eclipse for his first appearance on the world stage was, interestingly, not far off the predicted time that Nostradamus anticipated that a "King of Terror" would appear.

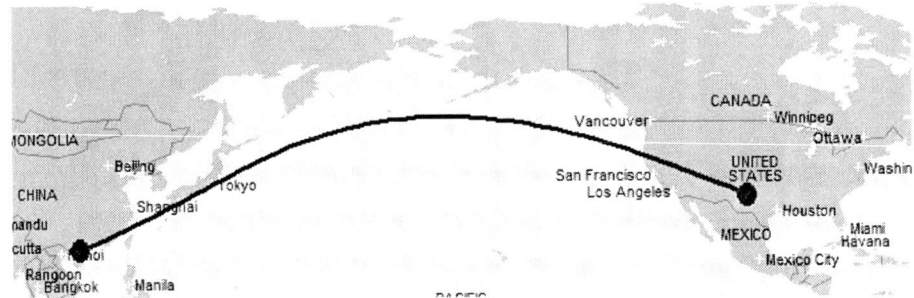

Putin returns as Russian President
China V USA Solar Eclipse of May 20, 2012

Two weeks after he assumed the office of president for the third time on May 7, 2012, the eclipse linking China to the USA via Japan occurred anticipating hostilities in the China seas. (For more on this current eclipse see page 127).

In many respects the Cold War eclipse of July, 1945, connecting Russia to the USA, although it occurred before his birth, also applies to him as his time in the KGB steeped him in Cold War attitudes.

Vladimir Putin has built up an anti-American alliance along the lines of these three eclipses. In the Middle East he has an alliance with Iran and in Asia with China. It seems probable that he will spearhead opposition to the USA and its NATO allies in combination with other enemies of America specified by the two eclipses of his ascendancies.

An advocate of a multi-polar geo-political world his Russian nationalism and global schismatic attitude could spill over into significant belligerence at a global level as the energies of these eclipses work through.

Putin's Formative Years
Cold War Solar Eclipse of July 9, 1945

You Must Be Joking

You would be wrong to think that eclipses don't have a sense of humour. In the UK, television comedy programmes such as Dad's Army, 'Allo, 'Allo and it ain't half hot mum have made light of the tragedies of eclipse events such as the Second World War and the British occupation of India. In the US the black humour of M.A.S.H., set in a military hospital in Korea during its war, pokes fun at life whilst all around is blood and gore; similarly the book Catch 22 set in the Mediterranean during World War 2. In August, 2006 the British comedian Sasha Baron Cohen released a film based upon one of his inner idiot characters, Borat, a television reporter from Kazakhstan. Why did Kazakhstan, an obscure central Asian former Soviet Republic, become the focal point of global humour at that time? The course of the solar eclipse of March 29, 2006, reveals why.

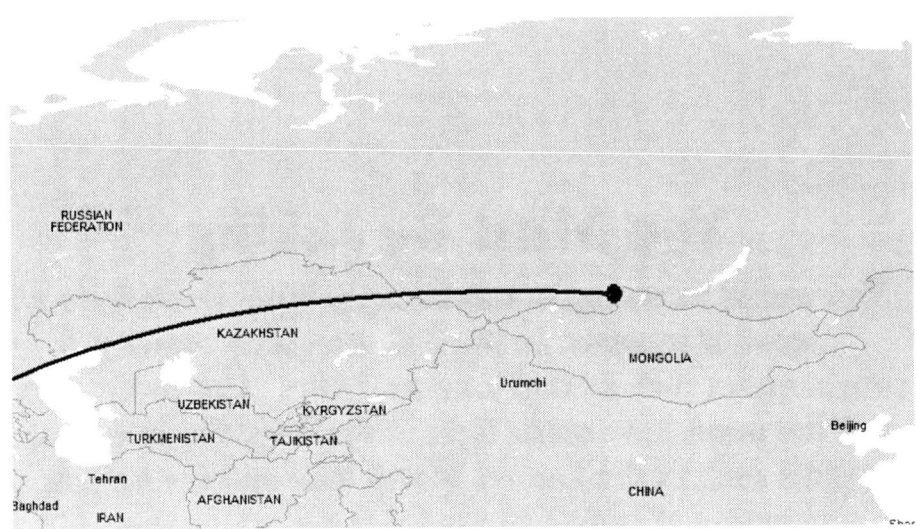

Borat's Kazakhstan
Solar Eclipse of March 29, 2006

If the other options are invasion, civil war or political fragmentation then being the subject of sniggering ridicule is very much a let-off to be grateful for.

Just as it was for Khazakstan when the film Borat was released and made it an international joke five months after this eclipse passed over it.

Borat:The Cultural Learnings of America for Make Benefit Glorious Nation of Khazakhstan was both very offensive and very hilarious incorporating deception, manipulation, farce and the exposure of prejudices.

Khazakstan's leaders were reportedly not amused but otherwise, thankfully, given the alternatives, experienced only hurt pride from their country's ignoble catapulting into the path of the comedic Zeitgeist.

Contemporary Solar Eclipses

Two recent eclipses stand out as the instigators of current regional conflict and power play. Their full impacts are still to be manifested but their effects are already apparent. They should also be taken in conjunction with the era-making NATO-Islamist eclipse of August 11, 1999.

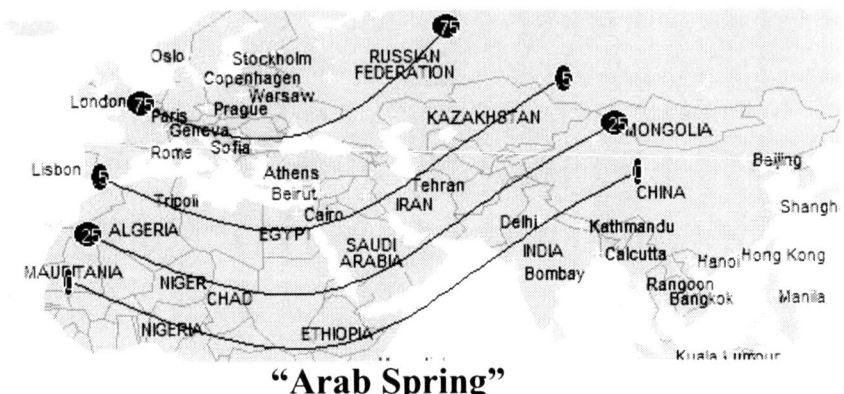

"Arab Spring"
Partial Eclipse of January 4, 2011

Mohamed Bouazizi, a street vendor of fruit and vegetables in the provincial Tunisian town of Sidi Bouzid, died in the afternoon of the same day as this eclipse. It would have been 50% partially visible when it passed over his hospital.

Eighteen days earlier he had poured gasoline over himself in the street outside the town's municipal offices and set himself alight protesting that he was unable to make a living because the authorities kept confiscating his cart and scales. Suffering 90% burns he was in a coma for the intervening period, his bandaged body giving him the appearance of an Egyptian mummy when he was photographed being visited by the country's president Zine El Abidene Ben Ali. His death first sparked riots and demonstrations in Tunisia leading to the departure of the president ten days later and then throughout the region including copy-cat self immolations. The Arab Spring had begun.

This eclipse was partially visible from North Africa and the Middle East. Since it occurred despotic governments have fallen in Tunisia, Libya, Egypt and Yemen and Syria is in the throes of a civil war. Two years after it street battles were still occurring in Egypt; 60,000 people had died in Syria and Israel was threatening to attack Iran. The political landscape has been transformed with democratic electorates voting in Islamist parties whose ties to the West are fractious to say the least.

In history it can be viewed as one of the most dramatic faultlines to have occurred in the region since the break-up of the Ottoman Empire.

Countries in Southern Europe were also caught in this eclipse and experienced significant changes to their democratically-elected governments owing to the Eurozone crisis and high government debt. These were carried out relatively peaceably although violent demonstrations did occur.

Timeline of the Wheals of the January 4, 2011, Eclipse

December 17, 2010   Tunisian street vendor Mohammad Bouazizi sets fire to himself in protest at the confiscation of his cart and scales.

January 4, 2011   Partial eclipse over Southern Europe, the Middle East and North Africa. Mohammad Bouazizi dies the same day and riots start in Tunisia. Protests spread to Lebanon, Oman, Yemen, Egypt, Syria and Morocco.

January 14, 2011   The government of Tunisia is overthrown following clashes with protestors in which dozens died.

January 25, 2011   Thousands gather in Tahir Square, Cairo in protest at the Egyptian government of Hosni Mubarak.

February 11, 2011   Hosni Mubarak resigns and power transferred to the Egyptian military.

February 14, 2011   "Day of Rage" protests in Bahrain.

February 15, 2011   Protests break out in Benghazi at the Libyan government of Muammar Gaddafi.

March 3, 2011   The former prime minister of Egypt, Ahmed Shafik, resigns.

March 15, 2011   Bahrain declares martial law.

March 17, 2011   UN declares no-fly zone over Libya.

March 18, 2011   52 protestors are killed by snipers at Sanaa University in the Yemen.

June 3, 2011   Failed assassination attempt on Yemen's president, Ali Abdullah Saleh.

July 31, 2011   Syrian tanks storm Hama killing at least 80.

August 20-28, 2011   Battle of Tripoli in Libya, effectively ending the Gadaffi government.

October 5, 2011   Russia and China veto the Western bloc's UN Security Council resolution to level sanctions against Syrian government.

October 20, 2011   Gaddafi captured and killed.

October 23, 20011   The Islamist Ennahda becomes the largest party after Tunisia's first democratic elections.

November 11, 2011   The democratically elected government of Greece falls to be replaced by caretaker government formed by former European Central Bank vice president.

November 11, 2011   Syria is suspended from Arab League.

November 12, 2011   The democratically-elected government of Italy falls to be replaced by caretaker government led by former European Commissioner Mario Monti.

December 21, 20011   The European Central Bank refinances Europe's banks with Euro489 billion loans to 500 banks.

February 1, 2012   A riot at Port Said football stadium in Egypt leaves 79 people dead and 1,000 injured.

February 3, 2012   The Syrian government begins an attack on the city of Homs.

February 27, 2012   The President of Yemen resigns.

June 24, 2012   Mohammed Morsi, the Muslim Brotherhood candidate wins Egyptian presidential electoral run-off.

July 15, 2012   The International Red Cross calls Syria's conflict a civil war.

November 14, 2012   Hamas military leader Ahmed Jabari killed in targeted Israeli air strike

November 15, 2012   Rockets fired from Gaza Strip hit apartment blocks in Tel Aviv.

December 21, 2012   Thousands riot in Egypt in vote over Islamist constitution

March 3, 2013   The Western powers step up "non lethal" aid to Syrian Rebels

March 6, 2013   Philippino United Nations soldiers captured by Islamist Martyrs Brigade on Golan Heights.

March 12, 2013   The British announce they may provide military support to Syrian rebels as humanitarian crisis grows.

April 18, 2013   Syrian president Bashar al-Assad warns the West that it is effectively sponsoring al-Qaeda terrorists against his Alawite regime and, should it fall, the consequences for it would be catastrophic.

Please see thewhealsofgod.com for updates.

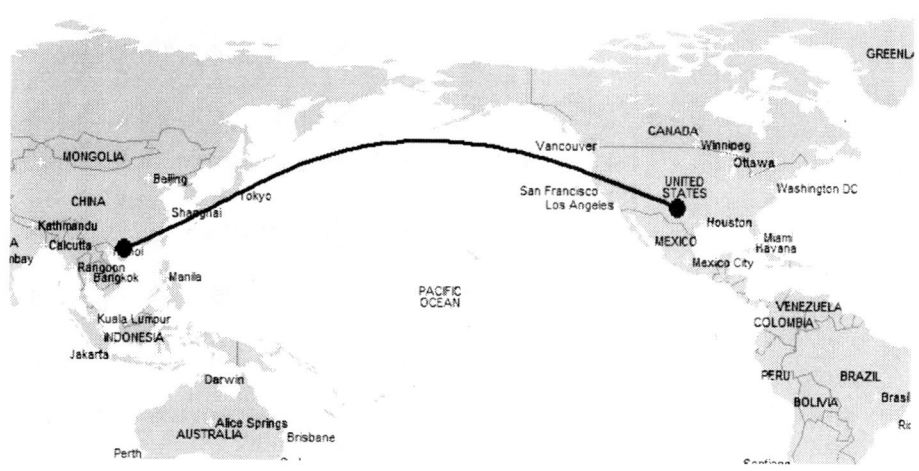

China V USA Solar Eclipse of May 20, 2012

This eclipse is of major geo-political significance at the present time as it highlights a potential conflict between the world's three largest economies, China, Japan and the USA as well as re-igniting the long running antagonism between North and South Korea.

In the run-up to it, the USA re-inforced its military presence in the region in response to China's growing naval power and more strident claims to islands in the East and South China seas.

Following it a breakdown

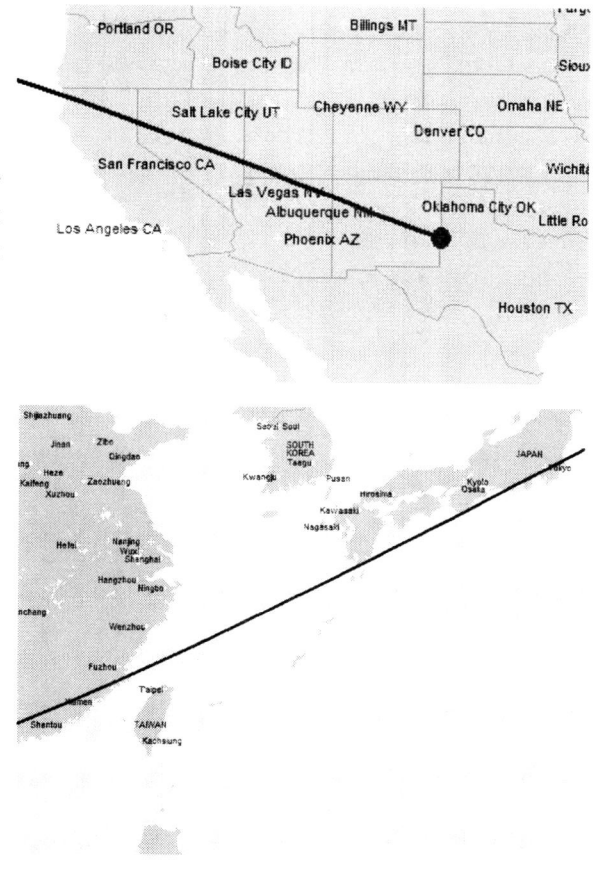

in the relationship between China and Japan began over the islands which saw anti-Japanese demonstrations in China including the destruction of Japanese-made vehicles.

These demonstrations aroused ancient wounds inflicted on China by the Japanese during its Imperialist period more than 100 years ago. Subsequently the Japanese have voted the Liberal Party back into power on a vehemently nationalist ticket adding fuel to the emotional sensitivities.

China has also alarmed other neighbours including Vietnam, the Philippines and Taiwan with a new map on its passports showing ownership of Taiwan and disputed South China Sea islands.

This eclipse is very much about a challenge to US power in the region from China with Japan acting as the catalyst.

The eclipse line also isolated South Korea making it more vulnerable to attack and conquest. Can the lid be put on these growing tensions between these countries on which so much of the world's economies depend?

Undoubtedly the USA will be drawn into any military dispute in the region if an escalation occurs. Arguably the Pacific has been an American Lake since the war with Spain in 1898 which saw them acquire the Philippines.

After the Second World War they have protected South Korea on the Asian mainland and Taiwan and Japan offshore although they were not able to hold onto South Vietnam.

China, however, for much of the past two centuries was prey to foreign powers especially those of Europe and Japan and can only recently be considered able of projecting naval power even into its adjacent waters.

This eclipse has several ominous features and much depends on whether hot heads can be cooled.

Amongst them is the overshadowing of both the location of the development of atomic weapons, Los Alamos, in New Mexico and the site where atomic weapons were first used, Hiroshima and Nagasaki in Japan.

Tanks, aerial bombers and aircraft carriers were developed towards the end of the First World War to become the chief weapons at the outset of the Second.

Similarly the weapons used at the end of the Second World War were long range missiles and atomic bombs could become the tools used to destroy the enemy in a Third World War.

Two years before the outbreak of the First World War the Krupps factory at Essen was overshadowed. Now it is the turn of the American military industrial complex in California and the south Western States of the USA including Arizona, New Mexico and Texas to be eclipsed.

Timeline of the Wheals of the May 20, 2012 Eclipse

<u>November 17, 2011</u>   US president Barack Obama announces a new military focus on Asia-Pacific whilst on tour in Australia to be known as "the Pivot." Commentators see it as challenge to China.

<u>April 19, 2012</u>   The nationalist right-wing mayor of Tokyo, Shintaro Ishihara, pledges to protect uninhabited islands disputed between Japan, China and Taiwan. Known as the Senkakus in Japan and Diaoyu in China they are surrounded by rich fishing grounds and potentially large natural gas deposits. China condemns the announcement saying it will take all necessary steps to control the islands.

<u>May 7, 2012</u>   Vladimir Putin is sworn in as president of Russia for his third non-consecutive term.

<u>May, 20, 2012</u>   The eclipse passes across the Pacific Rim beginning in the Gulf of Tonkin and ending in Texas via Eastern China, the East China Sea and Eastern Japan.

<u>June 6, 2012</u>   The Russian president Vladimir Putin reveals a military alliance with China including a stepping up of joint naval exercises in Asia Pacific. Widely viewed as a counter measure to the USA's increased presence and alliances in the region.

<u>August 14, 2012</u>   Seven Hong Kong activists reach the disputed islands of Senkaku/Diaoyu by sea.

<u>August 19, 2012</u>   Ten Japanese activists swim ashore and raise Japanese flag on the islands.

<u>September 11, 2012</u>   Japanese government announces that it has bought the East China Sea islands from their Japanese private owners. The islands, to the North of Taiwan, were passed over just South of the eclipse's line of totality.

<u>September 14, 2012</u>   Chinese surveillance ships step up pressure with a brief reconnoitre the islands.

<u>September 16, 2012</u>   Anti-Japan riots in China leading to the ransacking of Japanese businesses, destruction of Japanese cars and vandalisation of the Japanese embassy in Beijing. China reveals that it has submitted a document to the United Nations insisting that its continental shelf extends further than the standard 200 miles from its coastline justifying its claim to the islands.

<u>October 2, 2012</u>   Large-scale anti Chinese protests in Japanese cities.

<u>November 16, 2012</u>   Communist Party of China concludes once-in-a-decade CCP leadership transition with Xi Jinping and Le Keqiang assuming the mantles of Hu Jintao and Wen Jiabao.

<u>November 23, 2012</u>   China's new passports show map of islands in the East

and South China Seas as part of its territory provoking dispute with Taiwan, Vietnam, the Philippines, Brunei and Malaysia.

<u>December 26, 2012</u>   Shinzo Abe elected prime minister of Japan on a more nationalist and militarist ticket.

<u>January 18, 2013</u>   US Secretary of State, Hilary Clinton, says that USA has no position on the ultimate sovereignty of the islands but recognises that they are currently administered by Japan.

<u>February 7, 2013</u>   Russia denies its military jets strayed into Japanese airspace close to the four islands known as the Northern Territories, subject to dispute between Russia and Japan since their annexation by the Soviet Union at the end of World war II in 1945.

<u>February 8, 2013</u>   China denies that two of its ships put a radar lock on Japanese military vessels the previous month. Such lockings are usually a prelude to an attack.

<u>March 7, 2013</u>   North Korea re-asserts itself as a nuclear terrorist nation threatening the USA with use of first strike nuclear weapons as US and South Koreans carry out joint military manouevres.

<u>March 15, 2013</u>   The USA announces boost to Alaskan nuclear missile defences.

<u>March 20, 2013</u>   South Korean television stations and banks under cyber attack, generally held to be from North Korean sources.

March 28, 2013   US flies B-2 stealth bombers to South Korea as part of joint military manoeuvres.

<u>April 3, 2013</u>   North Korea closes down jointly owned factories close to border with South Korea.

<u>April 8, 2013</u>   North Korea warns foreigners to evacuate South Korea in case of war and positions missiles.

<u>April 9, 2013</u>   Patriot missiles placed around Tokyo to defend against possible North Korean attack.

<u>April 10, 2013</u>   China expresses its anger at deal between Japan and Taiwan over fishing rights around the disputed Senkaku/Diaoyu islands.

<u>April 13, 2013</u>   United States Secretary of State John Kerry visits his Chinese counterpart in Beijing to discuss ways of reducing tensions with North Korea.

Please see thewhealsofgod.com for updates.

Future Eclipses

It doesn't look good for the West in the coming years with both Europe, Western Russia and the USA in the firing line. Europe may well fragment if the implications of previous eclipses hold fast and it could be three and a half strikes and out for the USA in little over a decade. Central Africa, especially the Democratic Republic of Congo and the Indonesian archipelago are also in for testing times while conflict between Pakistan and India looks as though it will flare up again. China's bid for world hegemony looks likely to prove challenging for it from several directions including its economic imperialism in Africa, conquered territories such as Tibet and its domestic population. Also indicated is a strained relationship between Europe and Africa. Will a continental apartheid emerge?

European and Western Russian Collapse?
Eclipse of March 20, 2015

Occurring just ahead of the spring equinox, the birth point of the zodiac, this eclipse runs to the north of Europe and Russia and fits into a pattern established by similar encircling eclipse lines.

These include those around the American Civil War (1860), the Invasion of China by Japan (1936) and the collapse of the Soviet Union (1990). In the first case the political entity fought within itself; in the second it was invaded and in the third it fragmented.

If the EU is to implode economically or politically or be invaded militarily it would be around this time. Russia too could descend into civil war or a war of aggression concurrently, probably both.

This, of course, is a worst-case scenario and the eclipse may be more gentle with us although the already apparent stresses and strains within the European Union economically and politically and the tension between Russia and the West suggest that something has to give.

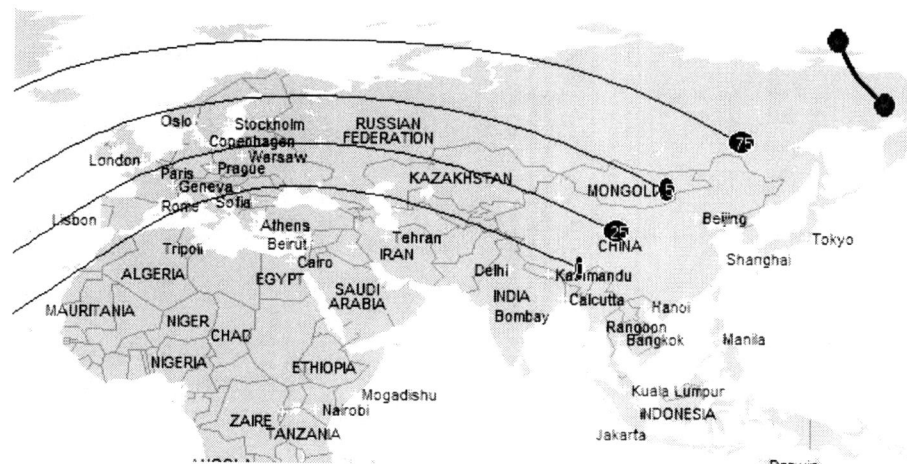

Europe and Russian travails continue
Partial Eclipses of June 10, 2021,
October 25, 2022 and March 29, 2025

A succession of partial eclipses, viewable from Europe and Western Russia in the first half of the 2020s, will re-inforce the impact of the 2015 eclipse.

Although partial eclipses are not to be considered as harmful as total or annular ones, the Jan 4, 2011 eclipse at the time of the Arab Spring, suggests that their impact can be significant if the conditions are

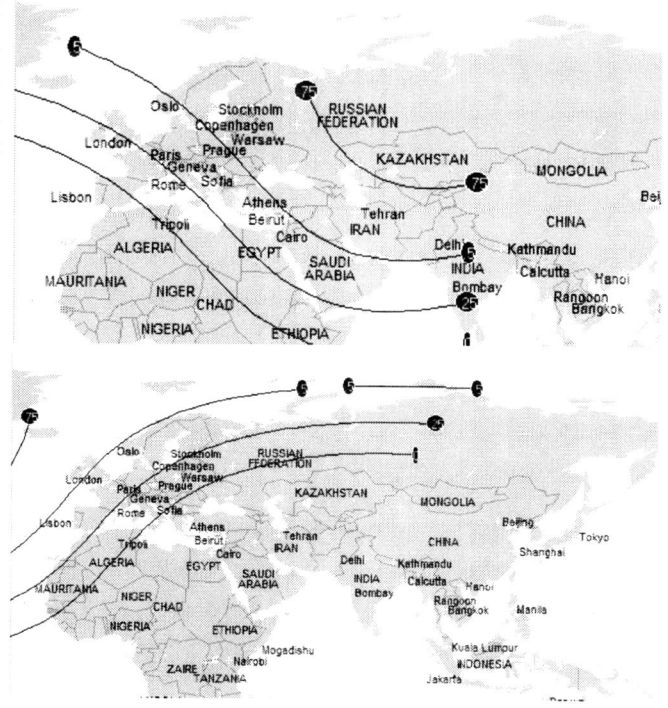

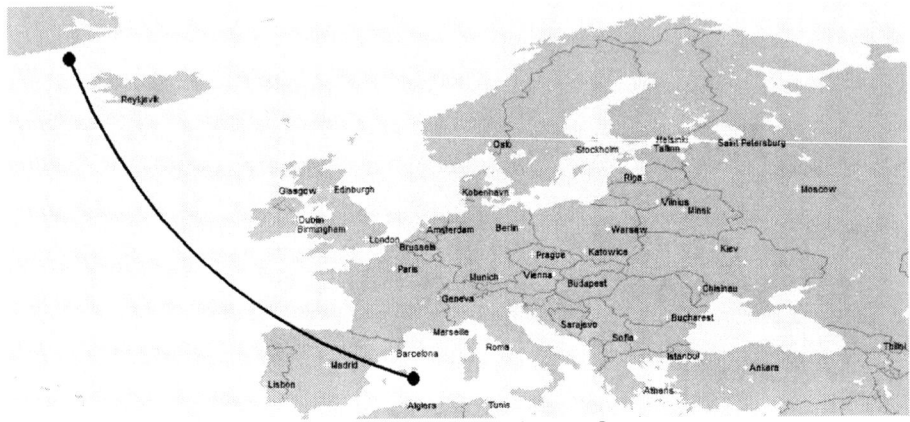

Europe Restricted?
Solar Eclipse of August 12, 2026

inflamed on the ground. The August 12, 2026, full eclipse falls in a rather strange way which cuts off Northern Europe from the North Atlantic. It gives the appearance of a hemming in or blockade.

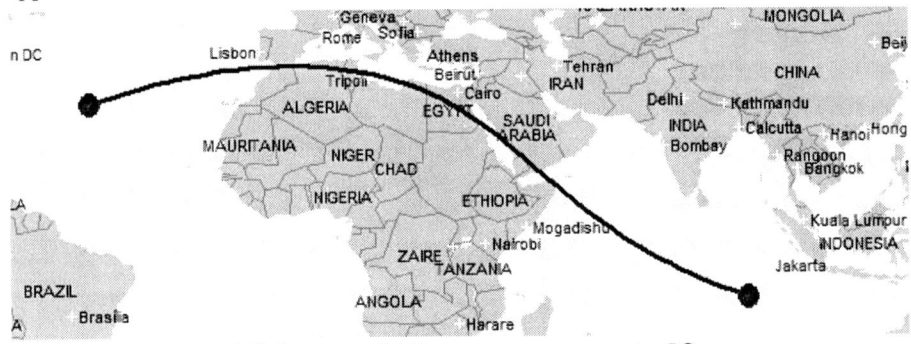

Africa and Europe separated?
Solar Eclipse of Aug 2, 2027

Shortly afterwards there appears what appears to be a definitive eclipse in the relationship between Europe and Africa. Past eclipses like this have seen power restricted between two continents such as the Monroe Doctrine Eclipse of 1821 which separated Europe from South America.

It could also be an encircling eclipse suggesting that Africa disintegrates into widespread conflict turning it, in a form of continental apartheid, into a No Go area, having shrugged off the post colonial nation state structure. We can see why this might arise in the next series of eclipses.

Democratic Republic of Congo (Zaire)

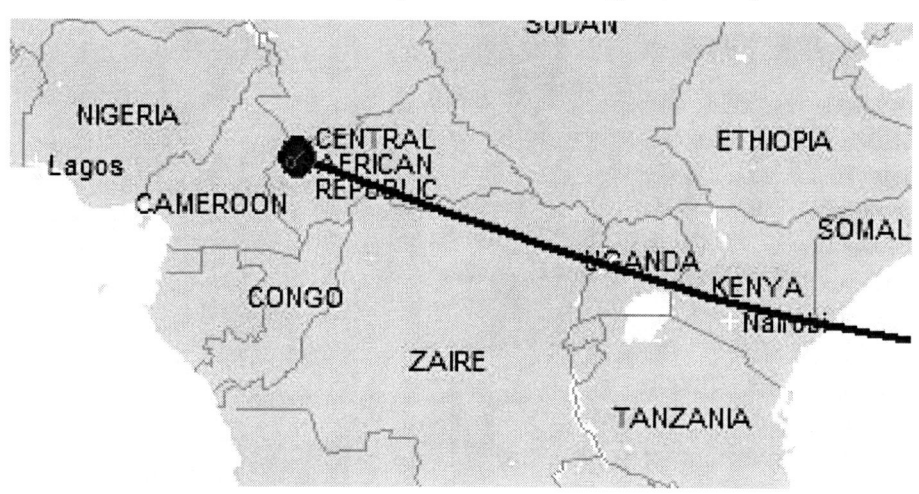

D. R. of Congo
Eclipse of January 15, 2010

D. R. of Congo
Eclipse of Nov 3, 2013

The prospects for Central and Eastern Africa, especially for the D. R. of Congo, also look rather bleak with a series of eclipses in this decade which are even more concentrated than those which afflicted Vietnam from the 1940s to the 1960s.

The D. R. of Congo receives an equatorial scouring five times in ten years, three of them in a space of three and a half years threatening Paraguayan levels of destruction.

Already engulfed in civil war and rebellion in the wake of the first of the

D. R. of Congo Eclipse of Sept 1, 2016

eclipses in 2010, the indications are that the conflicts can only escalate in their intensity. These eclipses will also affect its neighbours.

For example, the January 15, 2010 eclipse went over the Kenyan Somali border where significant fighting has broken out.

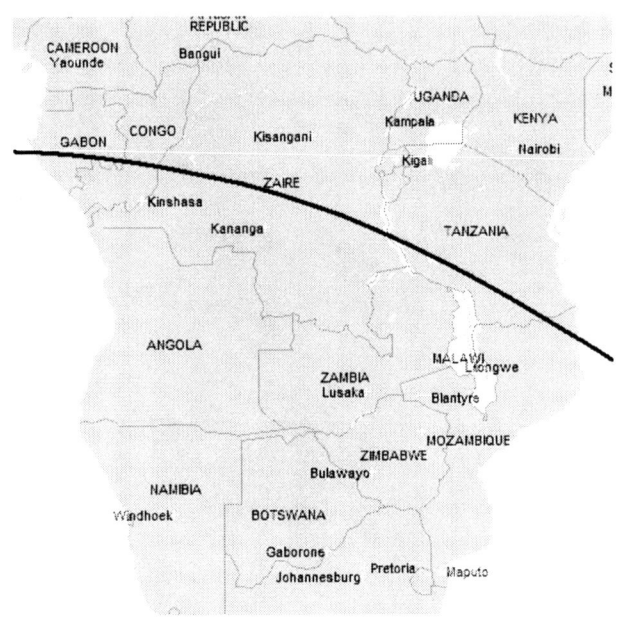

D. R. of Congo Eclipse of Feb 26, 2017

Uganda may also be embroiled in the near future as it receives a second eclipse in three years later this year. Rwanda and Burundi, already militarily active in Eastern D. R. of Congo are likely to reap a whirlwind.

These small former Belgian colonies have three successive close encounters of the eclipse kind coming up.

Angola and Sudan can

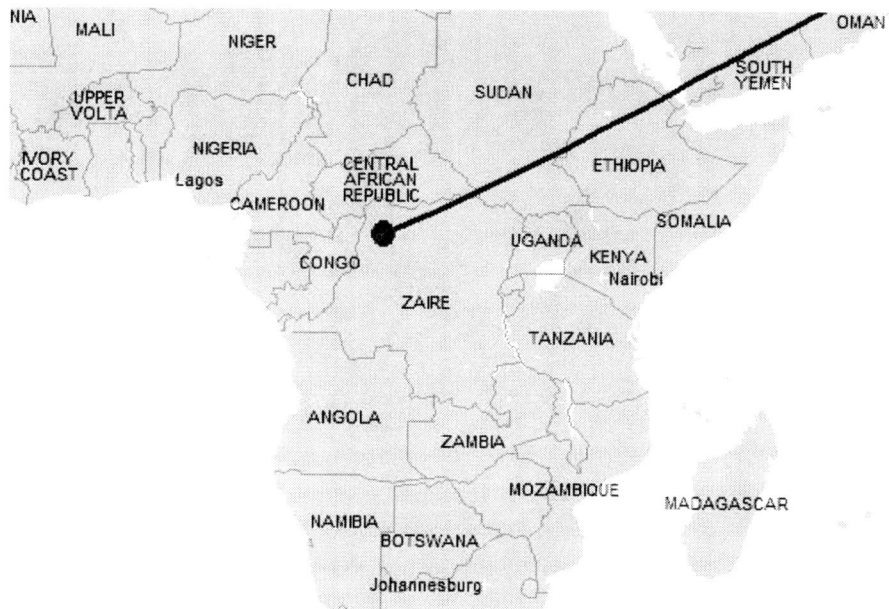

D. R. of Congo
Eclipse of June 21, 2020

expect more tribulations too, the former being divided by the 2017 and the latter by the one in 2020.

There will be a great deal of rumbles in the jungle in the coming years and D. R. of Congo is unlikely to remain a unified state at their end.

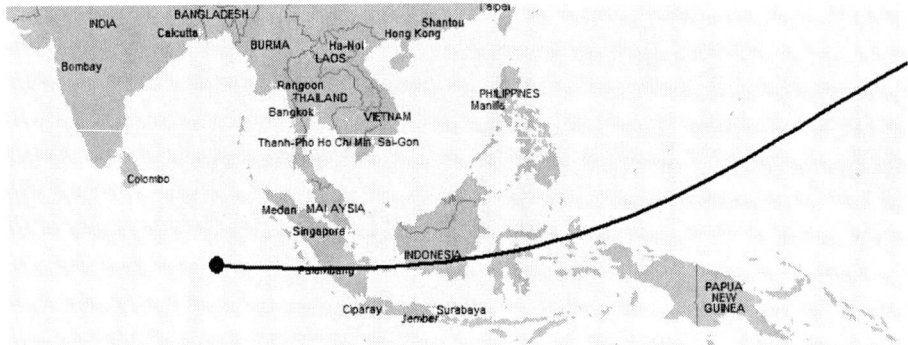

   With three eclipses in ten years criss-crossing its islands and two more over nearby Australia, the Indonesian archipelago, incorporating the states of Indonesia, Malaysia, Singapore, the Philippines and Papua New Guinea look as though they will become centres of agitation in the close future.

   Communal violence between various groups in Indonesia itself including Christians, Muslims, indigenous peoples, immigrants, Chinese and Malays have occurred frequently on the islands of the archipelago such as Sulawesi, West

Indonesia Under Pressure
Eclipses of March 8, 2016 and December 26, 2019

Additional Indonesian Eclipse
April 20, 2023

Timor, Java, Papua, Malaccas, Bali and Kalimantan and are likely to be inflamed.

Indonesia is a very large country inherited by tribal peoples after being brought together haphazardly by Dutch colonialists and it may struggle to survive intact following these eclipses.

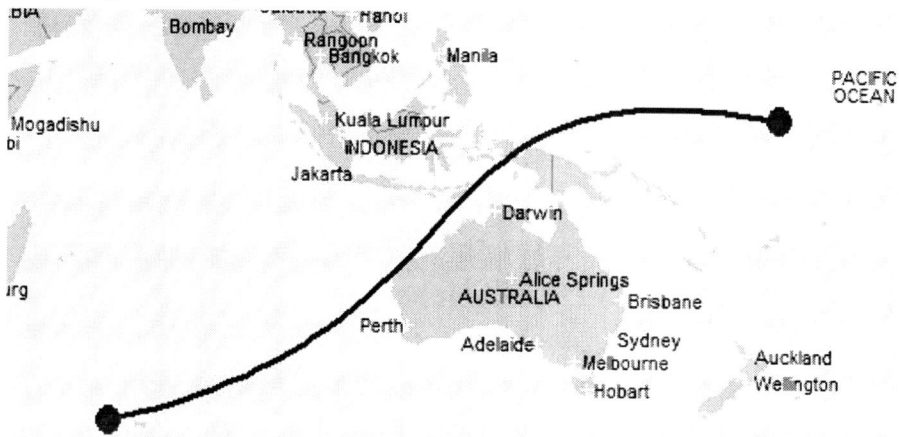

W(h)ither the USA?
Will the Flawed Superpower be Floored.

Just as Europe is due to experience a barrage of eclipses in the near future so too does the USA. The combined effects will undoubtedly place Mankind's gains through the West's development of science, technology, democracy and global governance at serious risk. The millennia-long efforts to make human life on earth pleasant, courteous and long is likely to be seriously challenged.

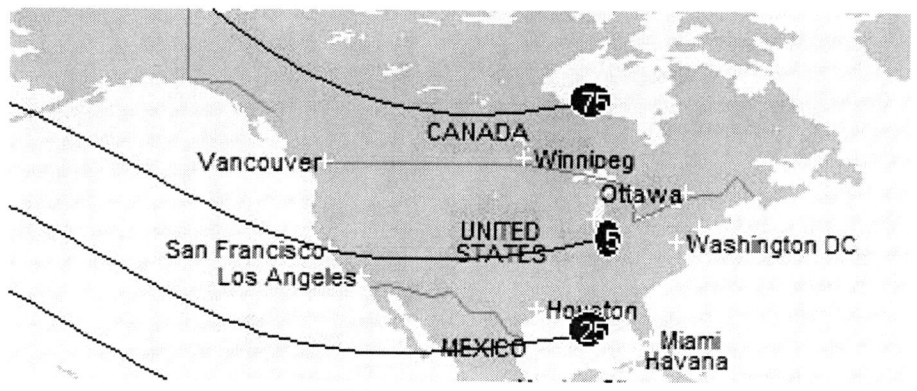

USA overshadowed
Partial eclipse of October 23, 2014

In a warm-up act for the harsher times that future eclipses will herald internal tensions are likely to bubble up closer to the time of this 50% occlusion over western and central USA.

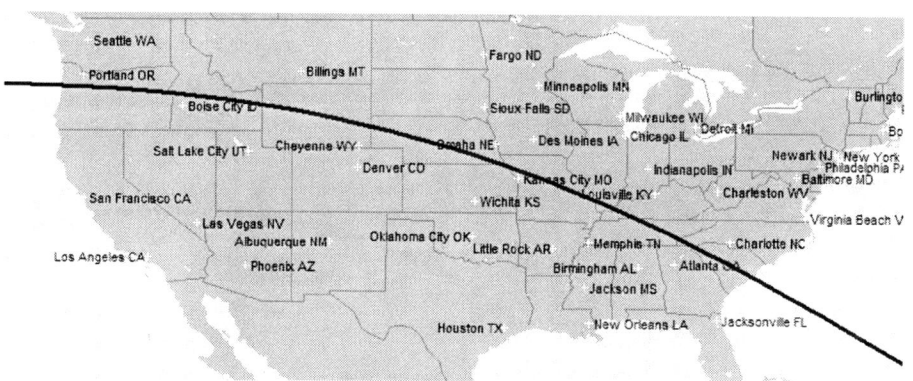

USA First Strike
Eclipse of August 17, 2017

This eclipse is a replica of the Oregon Trail eclipse in 1865 immediately after the American Civil War which sent the USA on a Westward course fighting native Americans and re-enslaving the newly-emancipated African Americans.

This could be the reversal of that experience with a significant showdown between the Obama coalition of newly-empowered American Blacks, Hispanics, students, Liberals and workers seeking a payback from the numerically shrinking but acquisitive WASP Tea Party elite.

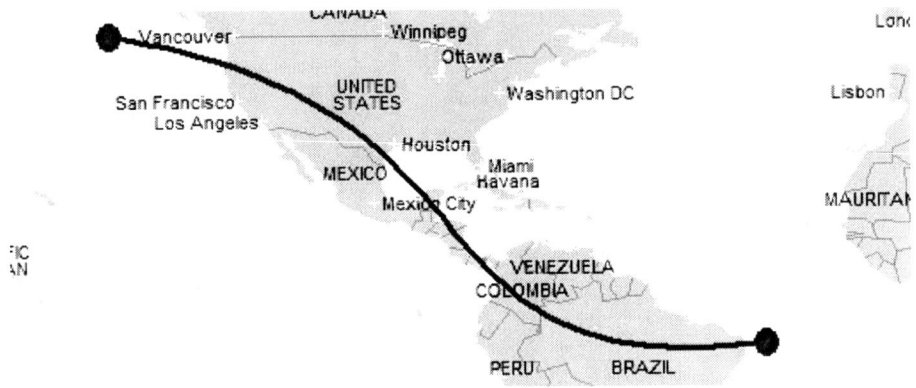

2nd USA Strike
Eclipse of October 14, 2023

Six years later the USA is to be struck by two more eclipses in succession. This first one follows a similar course to the May 20, 2012 eclipse through its South West states. Its vertiginous North-South line is redolent of the particularly strong eclipse of August 21, 1914 across eastern Europe which displaced many populations.

3rd USA strike
Eclipse of April 8, 2024

There will be little respite for the USA when a third total eclipse bisects it six months later, this time in an ominous south-north line comparable to the East Asian Communist line of 1948. The point where they cross is in Texas, not far from the end point of the May 20, 2012 eclipse. Texas has historically been in the line of fire of eclipses which probably accounts for its more common use of the death penalty.

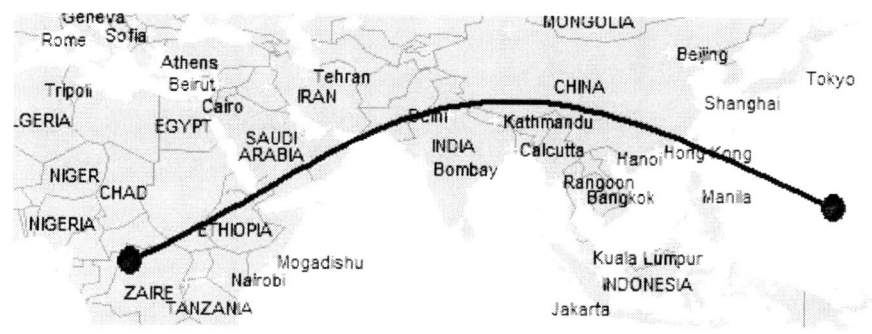

China, India and Pakistan Flashpoints
Eclipse of June 21, 2020

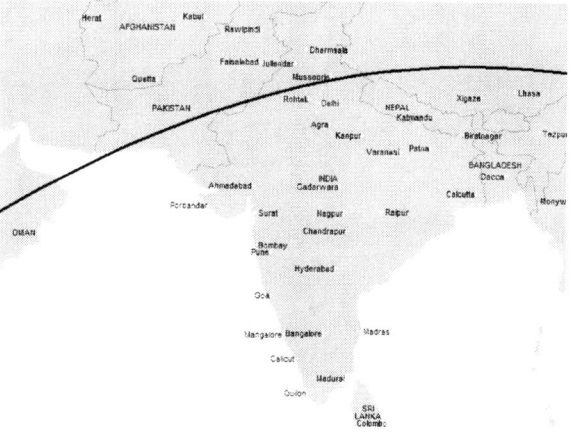

This eclipse will connect Africa to China, cross the India-Pakistan Border and ride overhead both Tibet and Taiwan.

It has echoes of the 1944 Independence eclipse but this one is more northerly and, essentially, is about China's reach.

Can China effect its influence in strategic parts of the world to ensure it is supplied with the minerals it requires for its development and manufacturing capabilities?

Will it be able to retain its economic ties to disputed Africa without resorting to arms?

Can it maintain its political control of rebellious Tibet without excessive violence?

And will it be confident enough to extend its power over recalcitrant and aggressively independent Taiwan?

It also passes over Chongqing, the central city of China's urbanisation and modernisation strategy. Can it be the superpower it wishes to be this eclipse will ask it?

India and Pakistan will be sorely tested too. Can they live peaceably together or will their entrenched animosities spill over into armed conflict?

Eclipses in Religion

As natural phenomena, eclipses are likely to be experienced, if only partially, at least once in nearly every natural lifetime so it is small wonder that perspectives on them have been incorporated into the world's great religions. The framers of these religions, themselves a product of the historical process, did not have the benefit of knowing the totality of where the eclipse paths fell apart from immediately overhead. They would not have been able to discern the conclusions I have made, being without the calculational and representational aids of modern computing. But they definitely had an inkling as to the true character of eclipses as this survey highlights.

Hinduism

In Hinduism the moon's nodes, which show where eclipses take place against the zodiac, have their own names, Rahu (North, Spring or Caput Draconis) and Ketu (South, Autumn or Cauda Draconis).

They are considered to be so influential that they have been given planetary status in horoscopic interpretation as well as their own periods of rulership of the life of the individual.

They are both considered to be malefic because they define the periods in a particular year when the light of the sun is occluded on the earth by the moon and also on the moon by the earth.

The deity image of Rahu is particularly telling. It is of the cut-off head of a demon snake. He is viewed as a deceiver, an anger divinity who plunges individuals and communities into chaos and cruelty.

Ketu is the remaining headless body of the snake. He causes sorrow and loss so that the individual turns to God and spiritualises himself. Ketu's effects can be likened to a crucifixion, bringing pain and troubles so that the material life has no joy.

These portrayals are derived from a myth concerning a demon who insinuated himself between the moon and the sun at a banquet to consume a nectar that bestowed immortality.

As it was being served the demon was detected by the sun and the moon, prompting Lord Vishnu to sever his head from his body to prevent him gaining eternal life.

However he was not quick enough, for the demon had already tasted it and became able to create upset whenever the sun and the moon were conjunct or opposite each other whilst near to either its head or tail.

There is a rich spirituality in this story, one which encompasses understandings of the demonic, the relationship between mortality and immortality and of 'being' in 'non-being'. The demon attains immortality at the moment of its death but is bisected and only capable of evil.

Hindus are highly sensitive to planetary movements and many of their festivals, prayers and devotions are based upon their cycles.

They consider that the days of eclipses are best for worship and the washing away of sins, especially through the reciting of mantras. Penance and good deeds on eclipse days are thought to have particular efficacy.

The great Hindu epic war poem, the Mahabharata, mentions three solar eclipses, including one which is a crucial determinant as a battle takes place.

Judaism

God is many things to the Jewish people, amongst them, as the historical process. The Bible tells their history which is viewed as the unfolding of God's will. The story of their escape from bondage in Egypt under the leadership of Moses in Exodus is central to their religion. Their study of history is, in part, so that they can know God, his mysterious ways, if you like, more exactly.

In many respects the title of this book acknowledges this understanding. But it takes it further by proposing that God's tool for manipulating the historical process is the eclipse. His will, past, present and future, can be known by the study of their timing and localities.

As we have seen eclipses occur around human events filled with suffering, occasionally around the appearance of total war. This connects to an aspect of God associated with one of the sephiroth on the Tree of Life in the Jewish Kabbalah. This is Geburah which stands for severity, strength and justice. It is like the planet Mars and has the Divine Name Elohim Gibor which means God of Armies.

The martial nature of eclipses also resonates with the original understanding of Jahweh by the Israelites as a God of War. According to Biblical scholarship Jahweh was introduced to the Israelites as a Storm God providing supernatural support for them in battle. He marched at their head and was represented by the Ark of the Covenant.

Clearly a powerful tribal warrior God, he evolved later into the God of Creation and subsequently into the God of Peace of Christianity.

Abraham, the founder of Judaism, was a Chaldean who were very active in interpreting the movements of the planets and were aware of the malefic effects of eclipses.

Arguably Abraham's knowledge of the harmful effects of combinations of planets and eclipses prompted him to seek refuge in obeisance to a Creator God, monotheism emerging as a direct re-action to the realisation that adverse astral effects are active in the world.

Kabbalah

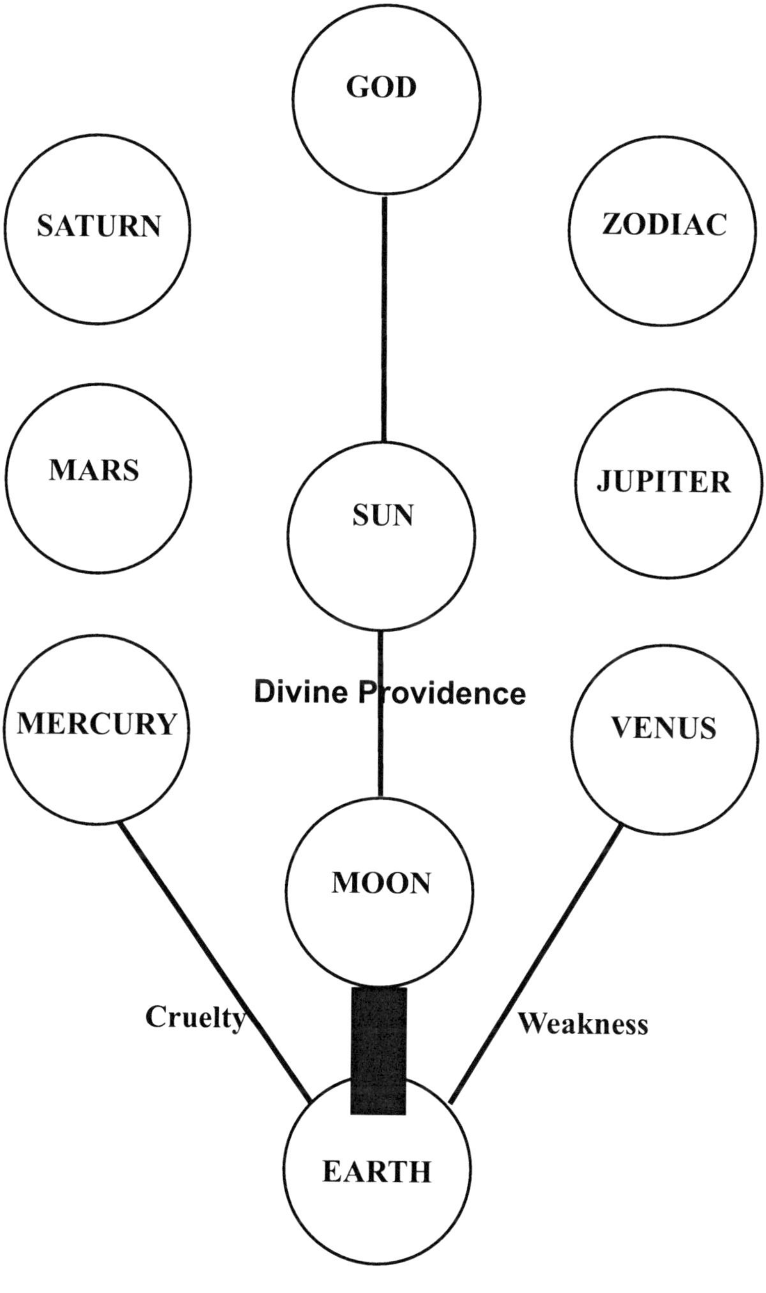

GOD

SATURN

ZODIAC

MARS

SUN

JUPITER

Divine Providence

MERCURY

VENUS

MOON

Cruelty

Weakness

EARTH

Kabbalists assign the ten sephiroth of the Tree of Life to God, the zodiac and to the earth and the seven traditional planets. Divided into three pillars, a core teaching is the attainment of the balance of the four sephiroth of the central pillar. These are assigned, in descending order, to God, the sun, the moon and to the earth. During an eclipse the light of the Providence of God as transmitted through the sun to the earth via the moon is physically blocked.

God's protective love cannot reach the earth through the middle pillar enabling the cruelty of the pillar of severity and the weakness of the pillar of mercy unrefined and unmoderated access to mankind.

The temporary imbalance caused by the loss of God's direct sustenance to humanity would, allegorically, account for the atrocities in history connected to eclipses.

A second kabbalistic way of understanding eclipses - as an interaction between the spiritual and material worlds - comes from the primary text of the kabbalah, the Sepher Yetzirah. The quote from it at the front of this book links the axis around which the universe revolves with the cycles of the sun, moon and planets and the heart of a human being. It outlines the relationship between universal space, cyclical time and moral action.

The axis is associated with the snake constellation of Draco which encircles the pole of the ecliptic and has stars in all 12 signs of the zodiac. In some texts is associated with the God Baal, who had a rival priesthood contending with the early Israelites notion of God. Modern Judaism appeals to God the Creator who is above the circumstances of the material universe which, allegorically, has a snake at its centre.

Solar eclipses occur in cycles timewise but appear to fall randomly in terms of the lines of shadow they make on the earth. The apparent arbitrary nature of these locations can now be viewed as agencies of the historical process such as the rise of the USA and China in modern times, the establishment and losses of the European empires and the globalisation of world politics. An order has been discerned in the chaos.

The heart of a human being is portrayed as the battleground between good and evil where the moral dimension of the universe is played out. In a number of ways nations and religions are similarly challenged by eclipses to see whether good or evil is paramount in the hearts of their decision-makers.

The hearts of national leaders are usually full of strong opinions and ambitions. They have strong egos and powers of assertion in order to compete for position with their peers and other nations, making it difficult for them, as Macchiavelli recognised, to put morality at the forefront of their pursuits.

Islam

Islamic writings on eclipses recognise their power for evil and punishment and propose they should be met by prayers to God as the creator of the sun and the moon for the length of their duration.

"When the sun was eclipsed at the time of the Prophet, he went out rushing nervously to the Mosque, dragging his cloak behind him, and led the people in prayer. He told them that the eclipse was one of the signs of Allah, with which Allah makes his slaves afraid and that it may be the cause of punishment coming on the people. He commanded them to do that which could prevent the punishment coming upon the people. So he commanded them to pray when an eclipse happens and to make du'aa (calls upon God), seek his forgiveness, give charity, free slaves, and do other righteous deeds so that the punishment would go away and not befall the people. So the eclipse is a reminder to people, making them afraid so that they will turn back to Allah and pay attention to him."

Mohammad was clear that he did not blame the death of his son Ibrahim on the day of an eclipse on the eclipse itself.

In some circles it is considered that the Islamic flag of a crescent moon and star is in fact an image of an eclipse in reference to the eclipse around the birth of Mohammad. Studies of the symbol have suggested that the shape of the partial disc is not that of a newish moon but rather of a partially occluded sun. Under the total path of an eclipse the stars become visible which may also be being indicated by the star incorporated into Islamic flags.

Although far from generally accepted some Christian scholars hold that Allah was a tribal Moon God so the image could be picking up on this association.

Some Islamic texts also point to the timing of the return of the Messiah or Mahdi being related to a particular combination of solar and lunar eclipses.

Buddhism

Buddhism is considered to be the distilled essence of Hinduism so it is not surprising that their myths about eclipses are similar, incorporating Rahu and Ketu. There are some interesting differences though, such as the involvement of urine.

The Tibetans recall that the Buddhas gathered to obtain the elixir of life, the antidote to the illnesses created by demons. Having obtained the elixir and placed it in the hands of a protector, Vajrapani, Rahu manages to steal it, drink it and then, being the deceptive trickster that he is, urinate it back into the pot.

Vajrapani then seeks Rahu out and slays him only to find that, having drunk the elixir, he comes back to life. As punishment for his failure to protect the elixir Vajrapani is forced to drink the urine. This enrages him against all demons, but especially Rahu whom he slays over and over again. Wherever Rahu's blood falls medicinal plants are said to grow.

Rahu seems to be portrayed as a contradiction, a slain immortal, one who dies at the moment he gains eternal life, a being who is also a no thing. Rahu is the more alive part of the dissected demon, the toiler and the gainer, whilst Ketu is the dead part, the loser and more detached from the world.

Vajrapani, the urine-drinking Rahu slayer, is considered the protector who controls the wrathful gza demons associated with mental disorders, epilepsy and negative celestial influences.

Buddhists believe that the subtle energies of the body are depleted during an eclipse and that the life of beings, both seen and unseen, are threatened. Effects during an eclipse are magnified so meritorious practices are particularly rewarded.

Shamanism

Many traditional peoples seek to appease or frighten off spirits when an eclipse takes place. Some believe that the spirits of the dead are trying to eat the sun and sing and dance and shout to frighten them away during the event.

Some American indian tribes saw it as a wolf or a frog trying to eat the sun and shout to scare it off whilst another saw it as a bear.

Perhaps it is possible to connect the human sacrifices practised by the Aztec, Mayan and Inca shamanic cultures of Mesoamerica and the Andes around eclipses to the insidious effects of eclipses over their settlements over millennia. In time an eclipse mentality became institutionalised in the ruling elite, stimulating them to practise large-scale, ceremonial public sacrifices.

Christianity

Christianity has little to say about eclipses per se, although the three-hour darkness that reportedly occurred at the crucifixion of the Messiah, is often associated with the natural dread of them.

Christianity's belief that the sufferings of Jesus saved the world does reflect a view that Man is in need of a rescuer from his plight on earth, which this book proposes is one of regular whippings from the cosmos. It could also be argued that the Christian personification of evil, Satan, is an aspect of eclipses.

Perhaps the Magi following the Star were really looking for a location under an eclipse path and this was subsequently changed to become less ominous. However a star shines whereas an eclipse darkens so this correlation seems unlikely. Certainly the image of the crucifixion of Jesus Christ resonates with the condition of humanity being periodically flayed by the solar system.

Greek and Egyptian

The Greeks held the view that eclipses were an indication that the Gods were angry and regarded them as a bad omen.

They also wrote plays portraying Man in a tragic light and formulated the concept that the world was governed by a craftsman God known as the demiurge, held to be malevolent by the Gnostics because of his association with the material world.

Plutarch in the Moralia V reported that Typhon/Seth in Greek and Egyptian myth were associated with the blood-red shadows of the earth on the moon during lunar eclipses. Seth was the Egyptian God of storms, desert and chaos, a usurper who killed and mutilated his own brother Osiris.

Typhon was a huge grotesque consisting of legs composed of snake coils, a feathered human trunk and dragon heads emerging from his arms and shoulders. He blew winds that destroyed crops and drowned sailors. He made hissing sounds of various animals but mainly of snakes.

The Pythagoreans also linked Typhon to a polygon of 56 sides, which is the number of Aubrey Holes at Stonehenge believed to have been used for the prediction of eclipses.

The quote from Homer's The Odyssey, at the front of this book suggests a strong belief in the sinister nature of eclipses.

Astrological

Astrologers have been astute in identifying the characteristics of eclipses. After all they started off by observing the heavens and they knew eclipses to be malefic by an analysis of their after-effects.

Astrologers do not know how astrology works in the physical realm, just that it does. They do not say that there is a direct cause only that the movements of the sun, moon and planets against the backdrop of the zodiac parallels an individuals character and activities. It is best described by the Hermetic dictum As Above, So Below which does not require a causal link, only a mirroring or synchronicity.

However in the case of solar eclipses there is an analysable physical impact on the earth, the umbra and penumbra, which can be made subject to scientific analysis.

Most astrologers today will analyse where eclipses fall in an horoscope in a personal reading and also where the ingress points fall for mundane predictions but this work is the first in-depth study of the impact on history as to where the astronomical shadows actually fell. As such it may be viewed as a bridge between 'real" and astrological understandings.

Eclipses and the I Ching

The I Ching or Chinese Book of Changes seems to have the clearest understanding of how eclipses work in history.

One of its earliest versions goes back 3,000 years. It tells of a solar eclipse which inspired Zhou King Wu to overthrow the Shang dynasty. King Wu apparently viewed the eclipse as an omen sent from Heaven to march against the Shang.

This account, which incorporates the themes of fighting, territorial change and dynastic fall, has been matched with a solar eclipse in 1070 BCE. Some of the I Ching hexagrams are connected to eclipses or at least to "darkenings" and usually incorporate advice such as forbearance, patience and the expectations that times will be more challenging.

Eclipses and the Tarot

The 22 major arcana of the tarot are said to have images which describe every situation in life. The card that stands out in regard to eclipses is the Tower which shows a tall building being struck by lightning, the occupants from its upper levels falling to the ground. The lightning bolt is said to be the will of God in striking down those presumptuous to believe that they could rise to the heavens.

The Tower card's connections to eclipses is additionally emphasised by its association with the planet of war, Mars.

The 9/11 destruction of the World Trade Centre which revealed to the world the conflict between the West and militant Islamic terrorists outlined by the 1999 eclipse, is the Tower card, passenger aircraft being substituted for the lightning bolt.

Freemasonry

Freemasonry is rooted in Plato's belief that rulers need to bring an enlightened and benevolent approach to their role. Its primary function is the production of "philosopher kings."

It does this through the promotion of the virtues and a charitable nature to its candidates whilst affording them the opportunity to learn how to address and administer public meetings.

Spiritually it takes its members on a journey from seeing the reflected shadows shimmering in Plato's cave to one of perceiving the pure divine light above ground.

As such, although it has nothing specifically to say about eclipses, it would naturally see them as diminishing because of their darkening effects.

Freemasonry does encourage the study of astronomy as one of the seven Liberal Arts required in the education of philosopher kings but not for the reasons I have advocated here.

When a Freemason dies he is said to depart this "sublunary" world for the Grand Lodge above, an indication that the framers of Masonic ritual appreciated the tainted nature of life on earth and the impact of the moon on our lives.

It is not generally appreciated that the twin towers of the New York World Trade Centre were a Masonic symbol indicating the need to combine the duality and interconnectedness of the material and spiritual realms in our daily lives.

The 9/11 attack was a direct assault on, not just the USA, but also Freemasonry which has done so much to create Western modernity including science, technology, separated constitutional powers, liberalism and democracy.

The original Grand Lodge, that of England, was announced two years after an eclipse covered London in 1715, sparking the First Jacobite Rebellion.

Freemasonry too, as an epoch-maker, is connected to the power circuit of eclipses with many of the battles for modernity especially in Europe and the Americas being led by its members.

Eclipses in Prophecy

The word divine has two meanings, that of God and also of fortune telling. Divination can be achieved through many and varied tools such as cards, tea leaves and dice but true diviners have a sixth sense, a second sight or a third eye. They often have an ability to communicate with spirits or possess a highly developed form of intuition.

Many spiritual organisations hold that their leaders must have attained the Gift of Prophecy as an indication of their preferment by God whilst in other more superstitious circles such capabilities are rejected, derided and feared.

There are many prophecies around the present time not least of which are those of Edgar Cayce who seems to have described the Wheals around the "Arab Spring" eclipse of January 4, 2011, precisely.

He is reportedly said to have foreseen a Third World War even before the end of the Second World War, apparently saying it would begin with strife in Libya, Egypt and Damascus.

His dream vision informed him that the Third World War would start with the flattening of Damascus, which some believe to be the world's first city. Current developments suggest that we are not far off that situation.

Whilst our subject is overwhelmingly about solar eclipses, in certain Jewish circles there is an interest in the meaning of a rare sequence of lunar eclipses which will be occurring in 2014-2015.

Full eclipses of the moon give it a reddy appearance often held to be like blood but more scientifically attributed to only infra-red light getting around the earth to colour the moon.

Much play is being made that these eclipsed moons will occur during these years on Jahweh's Feast Days of Passover and Tabernacles, a so-called tetrad of blood moons.

This is a rare astronomical co-incidence and has taken place at highly significant junctures in Jewish history including around their 1492 expulsion from Spain and the European discovery of the Americas by Christopher Columbus.

The next time such a tetrad took place was over 450 years later around the establishment of Israel in 1948 to be repeated again in relative quick succession by the 1967 retaking of Jerusalem following the Six Days War.

The Jews believe that they will come under a similar pressure in the coming years with a similar victory but that on this occasion there will be no excuse for the Messiah not to appear.

154

Religious Implications of the Wheals of God

Just as Darwin's Origin of Species and the establishment of geological time superceded the, albeit allegorical, view of Creation expounded in Genesis, so the beliefs of the various religions can be reviewed and then merged around this fresh understanding of eclipses in a new religious paradigm.

It is the thesis of this book that God constantly challenges man to deal with conflicts and changes of power relations. Eclipses, as has been demonstrated time and time again, are whips of negativity causing confusion, cruelty and combat in order for growth and transformation to occur. It is the breaking of eggs in order to make the omelette. It is essential that we prepare ourselves morally and organisationally to deal with the instabilities that they bring.

All religions contain some universal truths but are relative to the time of their formulation and to the partial group of humanity in which they emerged. Although all pertain to universality none has all of absolute truth within their body of knowledge.

Now that we have this new understanding of the astronomical basis of many afflictions on earth we have an opportunity to develop a form of cosmic humanism where we do not find our enemies in other human beings but in the very physics of our solar system and lunar satellite.

The focus of our activities needs to incorporate an awareness of humanity's shared astronomical situation, an easing away from our basic foci on economics, acquisition and status. It puts science at the forefront to try and understand how an eclipse can make people fight each other and bring about awful tolls of misery.

Scientists need to study the light and acoustic waves around eclipses to see how it might connect to the brain. Religious leaders need to incorporate this new appreciation of eclipses into the minds of their followers, disciples and congregations.

From each religion something can be taken, the imaginative mythology of the shamen, the Hindus and the Buddhists, the suffering of Christ and Buddha and the prayers of Mohammad.

So too the wisdom of the astrologers and the Chinese framers of the I Ching and the need for balance and moderation as incorporated into the kabbalah and the awareness of the interplay of the positive and negative or light and darkness of the Kabbalists, Freemasons and Daoists.

Unfortunately some elements of the monotheistic religions have been caught up in false Messianism which is only serving to exacerbate differences between them. Now they can unite in an understanding of how evil enters into the world.

Political Implications of the Wheals of God

The negative impact of eclipses indicate that Man's political units, especially in an age of globalisation, are constantly being threatened with division, break-up and invasion from astronomical flays that seek the weakest points along their paths to pit Man against Man.

If Mankind's political units are constantly being shaken by eclipses then larger countries won't work over time. Indeed our current system of globalised organisation may not either.

Many of the chapters in this book have traced the upsets that have occurred to the more sizeable, geographically squarer states such as the USA, Russia and China and to those that have long North-South borders such as Vietnam, the Sudan and South Africa. Purely by their geographic constitution they are more vulnerable to being eclipsed and therefore to instability.

Although eclipses did encourage expansion, especially in the earlier stages of our formation of a globalised system of nation states, today there is nowhere to go except to squabble over the seas, offshore islands and the frigid zones.

The only peaceable way forward is towards greater, more equally balanced manufacturing and exploitation of resources, a form of global planned economy to the extent that no nation is allowed to become over dominant. However the present system of national self interest and a winner-takes-all mentality in political leaders flies against this.

The longest-lasting continuous state in history is China, stretching back to the third century B.C.. But as we have seen, China has been constantly racked by rebellions, civil wars and warlordism. It has really been a civil society with a titular and weak head.

The Chinese experience of ruling large populations has concluded with strong, centralised government where the emphasis is on social order rather than on freedom of speech.

This is at odds with the natural disposition of Europeans and Americans towards greater freedoms. But the promotion of liberalism to the rest of the world has generated mixed results as recent experiences in the Middle East and North Africa have shown. Perhaps the attitude of one size fits all should be re-assessed.

Karl Popper's views on the Open Society and its Enemies, written in the wake of the Second World War, still apply as dogmatic populations relocate into the West.

The USA is about to be put under significant pressure by a partial and three total eclipses in slightly over a decade. It is doubtful it can survive in its present

156

form. Similarly a federated Europe looks untenable with the 2015 dissolution eclipse looming and other partial eclipses following.

The only way forward, if Mankind is to remain organised on a global basis, is through a global federated constitution with a world government exercising a monopoly on military power and the ability to redistribute global wealth and production.

Smaller states than some of those currently existing should be the natural political unit so that the unipolar view of the world represented by the United States during the presidency of George W. Bush and the present hostile multi-polar view of the Russians, Chinese and Iranians is superceded.

This world government should be aware of the power of eclipses to break up states such as those which have split Korea, Vietnam, the Sudan and Georgia. On this understanding Darfur should have its independence as should Abkhazia; Japan will never have Sakhalin back and China should accept Taiwan's permanent independence. It should be understood too that North Korea's aggression towards the South will continue until the boundary is moved closer to the eclipse line which ran east of Seoul.

Scientific and Educational Implications

The correlation of eclipses with the historical process should be understood and adopted by the scientific community as a matter of urgency. Scientists need to study the physical and electromagnetic effects of eclipses on the human brain with a view to counteracting them.

This book is potentially the seeding of a process which could end "scorched earth" wars on earth with the huge benefits that can bring especially as we now have such huge and interconnected populations.

Its core ideas should be taught in schools so that future generations can appreciate the situation they will be in as adult, decision-making, humanitarian, citizens. They should be encouraged to appreciate that we are all situated on a planet in the solar system with a very large satellite and that the interplay between the earth, sun and moon has serious consequences for our communal and personal well-being.

The stakes are very high. Unless scientists and political leaders can moderate the pernicious influences of eclipes on populations Mankind may not be able to sustain itself at its present beneficial level of trade, co-operation and technology.

Just as the Tower of Babel collapsed, Man's attempts to unify the world from a myriad of diverse ethnicities, cultures and religions into an humanitarian rainbow of peoples is constantly being threatened with descent into an horrendous Molotov Cocktail of nations.

157

The following table lists the 500 or so eclipses that will have occurred between 1801 and 2027. This study has been limited to the historical eclipse graphics available in Solar Fire's Solar Maps programme, which currently only go back to the beginning of the 19th century. It is sufficient however to establish the principles of the correlation of significant numbers of them to the world's past major geo-political events. The paths of future eclipses are available up to 2200 AD but it seems unnecessary to speculate past those of the next 14 years which suggest the river of history is entering a phase of rapids and waterfalls. As you will see many of our past eclipses are either of a remote or a small partial nature and have left and will leave no noticeable historical effect. The majority are local and their impact is not a core part of the world historical record although no doubt important for individual nations as the case study of the history of Sudan reveals. These have been marked as uninvestigated and remain subject to possible detailed analysis in the future.

Eclipse Correlation Table 1801-2027

ECLIPSE PATHS 1801-2027			
DATE	**PATH LOCATIONS**	**EVENT**	**TYPE**
Mar 14, 1801	Off South Africa.	Remote, Partial	Partial 19%
Apr 13, 1801	Central Russia, Turkey, Arabia, Iran, Kazakhstan	Uninvestigated	Partial 42%
Sept 8, 1801	Eastern Russia, off Japan Coast, Pacific	Remote, Partial	Partial 16%
Oct 7, 1801	New Zealand, Pacific, Antarctica	Remote, Partial	Partial 35%
Mar 4, 1802	Antarctica, Australia (Adelaide, Brisbane), Papua New Guinea, New Zealand	Uninvestigated	Central Total
Aug 28, 1802	Arctic, Central Russia, Mongolia, China, East China Sea.	**Beijing and Shanghai Eclipsed**	Central Annular
Feb 21, 1803	Off Australian Coast (Brisbane), Pacific, Mexico City, Miami.	**Louisiana Purchase**	Central Total
Aug 17, 1803	North Africa, Saudi Arabia, off Perth	**Mecca overrun by Saudis**	Central Annular
Feb 11, 1804	North Africa, Italy, Hapsburg Empire, Eastern Europe, Russia	**Austrian Empire proclaimed; Dissolution of Holy Roman Empire**	Central Annular
Aug 5, 1804	Pacific, Chile (Santiago) Argentina, Atlantic (Falklands)	Uninvestigated	Central Total
Jan 1, 1805	Off S. American coast, Antarctica	Remote, Partial	Partial 6%
Jan 30, 1805	Pacific, Oregon, Canada	Remote, Partial	Partial 17%
Jun 26, 1805	North East Asia, North America	Remote	Partial 94%
Dec 21, 1805	Antarctica	Remote	Central Annular
Jun 16, 1806	N. America, Atlantic, North Africa, Europe	Uninvestigated	Central Total
Dec 10, 1806	Australia, Pacific	Remote No Impact	Central Annular
Jun 6, 1807	Madagascar, Jakarta, Brisbane	Uninvestigated	Central Annular
Nov 29, 1807	Atlantic, North Central Africa (Nigeria), South Europe, Arabia.	Uninvestigated	Central Annular
May 25, 1808	Antarctica - South Africa	Remote, Partial	Partial 51%
Oct 19, 1808	Antarctica	Remote, Partial	Partial 17%
Nov 18, 1808	Mongolia, Japan	Uninvestigated	Partial 66%
Apr 14, 1809	On top of North America, US Canada in Shadow	Uninvestigated	Central Annular
Oct 9, 1809	South Atlantic, Antarctica	Remote	Central Total
Apr 4, 1810	Indonesia, Pacific	Uninvestigated	Central Annular
Sep 28, 1810	Pacific, Ecuador, Peru, Brazil, South Atlantic	**Start of South American Independence**	Central Annular
Mar 24, 1811	Tierra del Fuego, Falklands, Atlantic, Namibia, Zimbabwe	Uninvestigated	Central Total
Sep 17, 1811	Alaska, Canada, Great Lakes, US (Washington)	**British-US War**	Central Annular
Feb 12, 1812	British Columbia	Remote, Partial	Partial 34%
Mar 13, 1812	Antarctica	Remote, Partial	Partial 46%
Aug 7, 1812	Antarctica	Remote, Partial	Partial 23%
Sept 5, 1812	Arctic, Greenland	Remote, Partial	Partial 29%
Feb 13, 1813	North Africa, Europe, Russia	**Fall of Napoleon**	Central Annular

Date	Location	Event	Eclipse Type
Jul 27, 1813	Bolivia, Brazil	**South American Independence**	Central Total
Jan 21, 1814	Peru, Bolivia, Brazil	**South American Independence**	Central Annular
Jul 17, 1814	Chad, Arabia, Nepal, Philippines	Uninvestigated	Central Total
Jan 10, 1815	South Atlantic, Antarctica Southern America partially	Uninvestigated	Central Annular
Jul 6, 1815	Canada, Russia, Arctic	Uninvestigated	Central Total
Dec 30, 1815	Antarctica	Remote, Partial	Partial 43%
May 27, 1816	South of Australia	Remote	Central Annular
Nov 19, 1816	Poland, Romania, Turkey, Iran, India, China	Uninvestigated	Central Total
May 16, 1817	South Africa, Sri Lanka, Thailand, Philippines	Uninvestigated	Central Annular
Nov 9, 1817	Tibet, Hong Kong, Philippines, Papua New Guinea, Pacific	Uninvestigated	Central Total
May 5, 1818	Extends right across Russia!	**Karl Marx Born**	Central Annular
Oct 29, 1818	South Pacific, South Atlantic	Remote	Central Total
Mar 25, 1819	South Pacific, South Atlantic	Remote, Partial	Partial 13%
Apr 24, 1819	Atlantic, Arctic, Great Lakes	Remote, Partial	Partial 52%
Sept 19, 1819	Central Russia	Remote, Partial	Partial 6%
Oct 19, 1819	Antarctica	Remote, Partial	Partial 41%
Mar 14, 1820	Antarctica, Southern Africa, Madagascar	Uninvestigated	Central Total
Sept 7, 1820	Germany, Italy, Adriatic, Greece	**Greek Independence Revolt. Egypt expands into Sudan**	Central Annular
Mar 4, 1821	South Africa, Indonesia, Philippines	Uninvestigated	Central Total
Aug 21, 1821	Future Mexico-US Border, Texas, Southern States, South Atlantic off coast of Brazil and Angola	**Bisection of Mexican lands, Monroe Doctrine, Death of Napoleon**	Central Annular
Feb 21, 1822	North America, Canada, Baffin Island	Uninvestigated	Central Annular
Aug 16, 1822	Australia, Pacific	Uninvestigated	Central Total
Jan 12, 1823	Between New Zealand and Antarctica	Remote, Partial	Partial 5%
Feb 11, 1823	Russia	Remote, Partial	Partial 19%
Jul 8, 1823	North Russia	Remote, Partial	Partial 83%
Aug 6, 1823	Southern Argentina	Remote, Partial	Partial 27%
Jan 1, 1824	Antarctica	Remote	Central Annular
Jun 26, 1824	China, North America	Uninvestigated	Central Total
Dec 20, 1824	Brazil, Cape Town, Indian Ocean	**Brazil Independence**	Central Annular
Jun 16, 1825	Brazil, Angola, Harare	**Brazil Independence**	Central Annular
Dec 9, 1825	Mexico City	**Emperor flees Mexico**	Central Annular
Jun 5, 1826	Argentina, Chile	Uninvestigated	Partial 64%
Oct 31, 1826	Antarctica	Remote, Partial	Partial 12%
Nov 29, 1826	North Africa, Southern Europe	Uninvestigated	Partial 68%
Apr 26, 1827	Russia, Asia	Uninvestigated	Central Annular
Oct 27, 1827	Antarctica, Southern South America	Uninvestigated	Central Annular

Eclipse Correlation Table 1801-2027

Date	Location	Event	Type
Apr 14, 1828	South Atlantic, Cameroon, Sudan, Arabia, Pakistan, India, Burma, China	Uninvestigated	Central Annular
Oct 9, 1828	Philippines, South Pacific	Uninvestigated	Central Annular
April 3, 1829	New Zealand, South Pacific	Remote	Central Total
Sept 28, 1829	Russia, Japan, Pacific	Uninvestigated	Central Annular
Feb 23, 1830	Russia	Uninvestigated	Partial 31%
Mar 24, 1830	Antarctica, South Atlantic	Uninvestigated	Partial 52%
Aug 18, 1830	Falklands, Antarctica	Remote, Partial	Partial 12%
Sept 17, 1830	North Russia	Uninvestigated	Partial 39%
Feb 12, 1831	Pacific, Baja California, Texas, Washington, Atlantic	**Texan Colonists Revolt**	Central Annular
Aug 7, 1831	Australia, South Pacific	Uninvestigated	Central Total
Feb 31, 1832	Australia, Pacific, Mexico	**Texan Colonists Revolt**	Central Annular
Jul 27, 1832	Mexico, West Africa, Tanzania	**Texan Colonists Revolt**	Central Total
Jan 20, 1833	South Pacific, Argentina	Uninvestigated	Central Annular
Jul 17, 1833	Iceland, Arctic, Kamchatka	Remote No Impact	Central Total
Jan 9, 1834	Antarctica	Remote No Impact	Partial 44%
Jun 7, 1834	Southern Africa	**Queen Adelaide Province Annexed**	Partial 93%
Nov 30, 1834	Continental USA	**Texas-Mexico War**	Central Total
May 27, 1835	South America, Paraguay, West Africa, Nigeria, Kenya	Uninvestigated	Central Annular
Nov 25, 1835	West Africa, Atlantic, Angola	Uninvestigated	Central Total
May 15, 1836	Nicaragua, Jamaica, South Scotland, Denmark, Warsaw, Kiev, Caspian Sea	Uninvestigated	Central Annular
Nov 9, 1836	Perth, Australia, New Zealand, South Pacific	Uninvestigated	Central Total
Apr 5, 1837	Antarctica	Remote, Partial	Partial 6%
May 4, 1837	Pacific, Canada	Uninvestigated	Partial 64%
Oct 29, 1837	Argentina, Antarctic	Remote, Partial	Partial 45%
Mar 25, 1838	South Pacific, off South America	Remote, Partial	Central Total
Sep 18, 1838	Great Lakes, Washington	Uninvestigated	Central Total
Mar 15, 1839	Argentina, Brazil, Atlantic West Africa, Egypt	Uninvestigated	Central Total
Sept 7, 1839	Japan, Pacific	Uninvestigated	Central Annular
Mar 4, 1840	India, China, Russia	Uninvestigated	Central Annular
Aug 20, 1840	Africa, Angola, Tanzania, Antarctica	Uninvestigated	Central Total
Jan 22, 1841	Above Antarctica south of Africa	Remote, Partial	Partial 3%
Feb 21, 1841	Atlantic, Scotland	Remote, Partial	Partial 21%
Jul 18, 1841	Central Europe, East Russia	Uninvestigated	Partial 66%
Aug 16, 1841	New Zealand, South Pacific, Antarctica	Uninvestigated	Partial 41%
Jan 11, 1842	Antarctica Tierra del Fuego	Remote	Central Annular
Jul 8, 1842	Lisbon, Marseilles, Vienna, Kiev, Mongolia, Shanghai	**Hong Kong ceded to Britain.**	Central Total
Dec 31, 1842	South Pacific, Peru, Bolivia, Brazil	Uninvestigated	Central Annular
Jun 27, 1843	Mid-Pacific, Peru, Bolivia, Paraguay	Uninvestigated	Central Annular

Dec 21, 1843	Arabia, South India, Malaysian Peninsular, South China Sea, Philippines	Uninvestigated	Central Total
Jun 16, 1844	Australia, South Pacific	Uninvestigated	Partial 78%
Nov 10, 1844	South Pacific, Antarctica	Remote, Partial	Partial 9%
Dec 9, 1844	USA	Uninvestigated	Partial 69%
May 6, 1845	Europe, Russia, China	Uninvestigated	Central Annular
Oct 30, 1845	Antarctica, Australia, New Zealand, South Pacific	Uninvestigated	Central Annular
Apr 24, 1846	Pacific, Mexico, Miami, Atlanta, Algeria	**US-Mexican War**	Central Annular
Oct 20, 1846	Ghana, Nigeria, Zaire, Tanzania, Indian Ocean, Madagascar, Australia, Ayers Rock	Uninvestigated	Central Annular
Apr 15, 1847	Indian Ocean, Australia	Uninvestigated	Central Total
Oct 9, 1847	Plymouth, Paris, Munich, Istanbul, Baghdad, Dubai, Hyderabad, Rangoon, Hanoi.	**1848 European revolutionary upheavals, Communist Manifesto**	Central Annular
Mar 5, 1848	Hudson Bay, North Atlantic, Iceland	Remote, Partial	Partial 27%
Apr 3, 1848	South Pacific	Remote, Partial	Partial 58%
Aug 28, 1848	South Pacific	Remote, Partial	Partial 1%
Sep 27, 1848	Greenland, Northern Europe, Russia, Mongolia, China	Uninvestigated	Partial 49%
Feb 23, 1849	Central China, Japan, Alaska	**Taiping Rebellion**	Central Annular
Aug 18, 1849	Partial South Eastern Africa, Australia	Uninvestigated	Central Total
Feb 12, 1850	Angola, Zimbabwe, Mozambique, Indonesia, Philippines	Uninvestigated	Central Annular
Aug 7, 1850	Pacific. Partial in Japan and US	Uninvestigated	Central Total
Feb 1, 1851	South India, Antarctica, Melbourne, Sydney, New Zealand	Uninvestigated	Central Annular
July 28, 1851	Canada, Norway, Sweden, Baltic, Eastern Europe, Warsaw, Kiev, Black Sea, Caucasus, Georgia	Uninvestigated	Central Total
Jan 21, 1852	Antarctica	Remote, Partial	Partial 46%
Jun 17, 1852	South America	Uninvestigated	Partial 78%
Dec 11, 1852	Northern China, Japan	**Nien Rebellion. Perry opens up Japan**	Central Total
June 6, 1853	Pacific, Brazil	Uninvestigated	Central Annular
Nov 30, 1853	Pacific, Peru, Brazil	Uninvestigated	Central Total
May 26, 1854	US-Canadian border	**Sealing of US and Canadian Borders**	Central Annular
Nov 20, 1854	Brazil, South Atlantic, Indian Ocean	Uninvestigated	Central Annular
May 16, 1855	Russia	Uninvestigated	Partial 76%
Nov 9, 1855	Antarctica	Remote, Partial	Partial 49%
April 5, 1856	Australia, Brisbane	Uninvestigated	Central Total
Sept 29, 1856	Western Pacific, Eastern Russia	Uninvestigated	Central Annular
Mar 25, 1857	Sydney, Mexico	Uninvestigated	Central Total
Sept 18, 1857	Ottoman Empire, North India	**Indian Mutiny**	Central Annular
Mar 15, 1858	Caracas, Atlantic, England, Oslo, Russia	**Establishment of Fenianism**	Central Annular
Sept 7, 1858	South America	Uninvestigated	Central Total
Feb 3, 1859	South Pacific	Remote, Partial	Partial 1%
March 4, 1859	Pacific, Western Canada	Remote, Partial	Partial 25%

Eclipse Correlation Table 1801-2027

Date	Location	Correlation	Type
July 29, 1859	North Russia, Canada	Remote, Partial	Partial 52%
Aug 28, 1859	South Indian Ocean, Antarctica	Remote	Partial 53%
Jan 23, 1860	South Pacific, Antarctica	Remote	Central Annular
July 18, 1860	Canada Atlantic, Spain, North Africa, Ethiopia	**American Civil War**	Central Total
Jan 11, 1861	Indian Ocean, Australia, Pacific	Uninvestigated	Central Annular
Jul 8, 1861	Indian Ocean, Malaysia, Philippines, Pacific	Uninvestigated	Central Annular
Dec 31, 1861	Caribbean, Atlantic, North Africa	Uninvestigated	Central Total
Jun 27, 1862	Indian Ocean, South Africa, Western Australia	Uninvestigated	Partial 92%
Nov 21, 1862	Antarctica	Remote, Partial	Partial 6%
Dec 21. 1862	The Stans, China	Uninvestigated	Partial 70%
May 17, 1863	Canada, Western USA, Europe	Uninvestigated	Partial 86%
Nov 11, 1863	Antarctica, Indian Ocean, South Africa, Western Australia	Remote	Central Annular
May 6, 1864	Pacific, Indonesia, Philippines, off Mexico	Uninvestigated	Central Annular
Oct 30, 1864	Pacific, South America, direct over Asuncion North Chile, Bolivia, Paraguay, South Brazil, Atlantic	**War of Triple Alliance**	Central Annular
April 25, 1865	South Pacific South America, Chile, Argentina, Uruguay, Brazil, Atlantic, Angola, Zambia.	**War of Triple Alliance**	Central Total
Oct 19,1865	USA - diagonal from Washington State to Georgia Coast, Atlantic, West Africa.	**Re-enslavement of Black Americans. Western Indian wars.**	Central Annular
Mar 16, 1866	North Pacific	Remote, Partial	Partial 21%
April 15, 1866	Antarctica, Indian Ocean, Australia	Remote	Partial 66%
Oct 8, 1866	Canada, Greenland, Atlantic Scotland	Uninvestigated	Partial 57%
Mar 6, 1867	Italy, Balkans, Moscow, Russia	**Dual Monarchy**	Central Annular
Aug 29, 1867	South America, Chile, Argentina, Buenos Aires, South Atlantic, Antarctica	Uninvestigated	Central Total
Feb 23, 1868	South America, Peru, Brazil, West Africa, Sudan.	Uninvestigated	Central Annular
Aug 18, 1868	Ethiopia, India, Malaysia, Indonesia	Uninvestigated	Central Total
Feb 11, 1869	South Atlantic, Cape Town	**South Africa's Diamond Rush**	Central Annular
Aug 7, 1869	Mongolia, Russia, Alaska, Canada, US, Washington	Alaska Purchase?	Central Total
Jan 31, 1870	Antarctica, South Atlantic	Remote, Partial	Partial 48%
Jun 28, 1870	South Pacific, New Zealand	Remote	Partial 63%
July 28, 1870	North Russia	Remote, Partial	Partial 7%

163

Date	Location	Event	Eclipse Type
Dec 22, 1870	Europe - Portugal, Sicily, Greece, Istanbul, Black Sea, Crimea	**Unification of Germany, Franco Prussian war.**	Central Total
Jun 18, 1871	Indonesia, Australia	Uninvestigated	Central Annular
Dec 12, 1871	India, Australia, Indonesia	Uninvestigated	Central Total
Jun 6, 1872	India, Burma, China, Korea, Japan	Uninvestigated	Central Annular
Nov 30, 1872	New Zealand, South Pacific, Southern South America, Chile, Argentina, Atlantic	Uninvestigated	Central Annular
May 26, 1873	Europe, Russia	Uninvestigated	Partial 90%
Nov 20, 1873	Antarctica	Remote	Partial 51%
Apr 16, 1874	Antarctica, South Africa	Uninvestigated	Central Total
Oct 10, 1874	Russia, Europe, Iran, Arabia, Ethiopia, Egypt	Uninvestigated	Central Annular
Apr 6, 1875	Cape Town, Indian Ocean, Thailand, Vietnam, Pacific	**Annexation of Transvaal**	Central Total
Sep 29, 1875	New York State, Atlantic, West Africa, Angola, Zaire	Uninvestigated	Central Annular
Mar 25, 1876	Pacific, British Columbia, Canada, Greenland	Uninvestigated	Central Annular
Sep 17, 1876	South Pacific	Remote	Central Total
Mar 15, 1877	Central Asia	Uninvestigated	Partial 29%
Aug 9, 1877	Kamchatka	Remote, Partial	Partial 33%
Sep 7, 1877	South America, Bolivia, Argentina, Chile	Uninvestigated	Partial 64%
Feb 2, 1878	Southern Indian Ocean, Antarctica	Uninvestigated	Central Annular
Jul 29, 1878	Mongolia, Russia, Alaska, Canada, US, Cuba	Repeat of Alaska Purchase	Central Total
Jan 22.1879	Southern Africa	**Zulu Wars**	Central Annular
Jul 19, 1879	West Africa, Guinea, Chad, Mali Sudan, Somalia	**Mahdi Revolt**	Central Annular
Jan 11, 1880	Pacific, West Coast US	**Wild West. Billy the Kid, OK Corral**	Central Total
July 7, 1880	South Atlantic, Southern South America	Uninvestigated	Central Annular
Dec 2, 1880	Antarctica	Remote, Partial	Partial 4%
Dec 31, 1880	Eastern US, Atlantic, UK, West Europe	Uninvestigated	Partial 71%
May 27, 1881	Canada, North Eastern Russia	Uninvestigated	Partial 74%
Nov 21, 1881	Antarctica, Southern South America	Uninvestigated	Central Annular
May 17, 1882	Nigeria, Egypt, Iran, China	**Sudanese War**	Central Total
Nov 10, 1882	Indonesia, Pacific	Uninvestigated	Central Annular
May 6, 1883	Pacific	Uninvestigated	Central Total
Oct 30, 1883	Pacific	Uninvestigated	Central Annular
Mar 27, 1884	Europe, Russia	Partial	Partial 14%
Apr 25, 1884	South Atlantic	Remote, Partial	Partial 76%
Oct 19, 1884	North Pacific	Remote	Partial 64%
Mar 16, 1885	Pacific, North West USA, Canada, Greenland, North Atlantic	Uninvestigated	Central Annular
Sep 8, 1885	New Zealand, South Pacific	Uninvestigated	Central Total

164

Eclipse Correlation Table 1801-2027

Mar 5, 1886	Papua New Guinea, Pacific, Mexico	Uninvestigated	Central Annular
Aug 29, 1886	Off North coast of South America, Atlantic, Angola, Zimbabwe	Uninvestigated	Central Total
Feb 22, 1887	Pacific, New Zealand, South American Coast	Uninvestigated	Central Annular
Aug 19, 1887	Berlin, Moscow, Tokyo	**Seeding of 20th Century wars**	Central Total
Feb 11, 1888	South Pacific	Remote No Impact	Partial 50%
Jul 9, 1888	South Pacific	Remote No Impact	Partial 48%
Aug 7, 1888	Greenland, Arctic	Remote No Impact	Partial 20%
Jan 1, 1889	Pacific, San Francisco, Montana, Winnipeg	Uninvestigated	Central Total
Jun 28, 1889	Southern Africa, Namibia, Botswana, Zambia, Malawi, Mozambique	**Scramble for Africa Rhodesia campaign**	Central Annular
Dec 22, 1889	Off Brazilian Coast, Angola, Zaire, Kenya, Somalia	**Scramble for Africa Uganda Campaign**	Central Total
17-Jun-90	North Africa, Europe, Asia, Senegal, Algeria, Libya, Turkey, Iran, India, Burma	**Scramble for Africa French Campaign into North West Africa**	Central Annular
Dec 12, 1890	Indian Ocean, Australia, Pacific Ocean	Uninvestigated	Central Annular
Jun 6, 1891	North America, Europe	Uninvestigated	Central Annular
Dec 1, 1891	Antarctica	Remote No Impact	Partial 53%
April 26, 1892	South Pacific, New Zealand	Remote No Impact	Central Total
Oct 20, 1892	North and Central America	Uninvestigated	Partial 90%
April 16, 1893	Chile, Argentina, Paraguay, Brazil, Atlantic, Senegal, Mali, Niger, Chad, Sudan (Khartoum)	**Scramble for (French West) Africa. Khalifa declines**	Central Total
Oct 9, 1893	Pacific, Peru	Uninvestigated	Central Annular
April 6, 1894	Indian Ocean, India, China, East Russia, Alaska	First Sino-Japanese War	Central Annular
Sep 29, 1894	Uganda	**Ugandan Protectorate created**	Central Total
Mar 26, 1895	North Atlantic	Remote, Partial	Partial 35%
Aug 20, 1895	Arctic, Urals	Remote, Partial	Partial 27%
Sept 18, 1895	South Pacific	Remote	Partial 74%
Feb 13, 1896	South Atlantic	Remote	Central Annular
Aug 9, 1896	North Norwegian Coast, North Russia, Eastern Europe, Kazakhstan	Uninvestigated	Central Total
Feb 1, 1897	New Zealand, Pacific, Colombia, Venezuela	Uninvestigated	Central Annular
July 29, 1897	North Mexico, Havana, Caribbean, Brazilian Coast, South Atlantic	**US-Spanish War**	Central Annular
Jan 22, 1898	Nigeria, Ethiopia, India, China	**Fashoda, Boxer Rebellion**	Central Total
Jul 18, 1898	South Pacific	Remote	Central Annular
Dec 13, 1898	Antarctica	Remote, Partial	Partial 2%
Jan 11, 1899	North Pacific	Remote	Partial 72%
Jun 8, 1899	North Russia	Remote	Partial 61%

Date	Location	Status	Type
Dec 3, 1899	South Atlantic, Antarctica	Remote	Central Annular
May 20, 1900	Mexico, New Orleans, Carolinas, Atlantic, Spain, Tunisia, Egypt	Uninvestigated	Central Total
Nov 22, 1900	Atlantic, Angola, Zambia, Indian Ocean, Western Australia	**Boer War Kitchener Escalation**	Central Annular
May 18, 1901	Madagascar, Indian Ocean, Indonesia	Uninvestigated	Central Total
Nov 11, 1901	Sicily, Cairo, Arabia, Sri Lanka, Cambodia, Vietnam	Uninvestigated	Central Annular
Apr 8, 1902	North West Canada	Remote, Partial	Partial 6%
May 7, 1902	South Pacific, New Zealand	Uninvestigated	Partial 86%
Oct 31, 1902	Europe, Russia, China	Uninvestigated	Partial 70%
Mar 29, 1903	China, Mongolia, Kamchatka, Arctic	Uninvestigated	Central Annular
Sep 21, 1903	South Indian Ocean	Remote	Central Total
Mar 17, 1904	Tanzania, Indian Ocean, Malay Peninsular, Vietnam, Philippines, Pacific	Uninvestigated	Central Annular
Sep 9, 1904	Pacific, Hawaii, Chile	Uninvestigated	Central Total
Mar 6, 1605	South Indian Ocean, Australia	Uninvestigated	Central Annular
Aug 30, 1905	Winnipeg, Atlantic, Spain, Tunisia, Libya, Egypt, Saudi Arabia	Uninvestigated	Central Total
Feb 23, 1905	Antarctica, South West Australia	Uninvestigated	Partial 54%
Jul 21, 1906	South Atlantic	Remote, Partial	Partial 34%
Aug 20, 1906	North Canada	Uninvestigated	Partial 51%
Jan 14, 1907	Russia, Stans, China, Russia	Uninvestigated	Central Total
Jul 10, 1907	Pacific, South America, Chile, Bolivia, Brazil, Atlantic	Uninvestigated	Central Annular
Jan 3, 1908	Pacific, Off Costa Rica	Uninvestigated	Central Total
Jun 28, 1908	Pacific, Mexico City, Caribbean, Florida, Atlantic, Senegal, Ghana	Uninvestigated	Central Annular
Dec 23, 1908	Chile, Argentina, South Atlantic	Uninvestigated	Central Annular
Jun 17, 1909	North America, Eastern Russia, China	Uninvestigated	Central Annular
Dec 12, 1909	South Atlantic, South Pacific	Remote	Partial 54%
May 10, 1909	Australia	Uninvestigated	Central Total
Nov 2, 1910	Kamchatka, Pacific	Uninvestigated	Partial 85%
Apr 28, 1911	New South Wales, Pacific, Off El Salvador coast.	Uninvestigated	Central Total
Oct 22, 1911	Stans, China, Philippines, Indonesia	**Fall of China's Emperor**	Central Annular
Apr 17, 1912	Europe - Portugal, Spain, Denmark, Baltic, Riga, North Russia	**First World War Western Front**	Central Annular
Oct 10, 1912	Galapagos, Columbia-Ecuador Border, Brazil, Atlantic	Uninvestigated	Central Total
Apr 6, 1913	Canada	Remote, Partial	Partial 42%
Aug 13, 1913	Eastern Canada	Remote, Partial	Partial 15%
Sep 30, 1913	South Indian Ocean	Remote	Partial 82%
Feb 25, 1914	Pacific	Remote	Central Annular

Eclipse Correlation Table 1801-2027

Aug 21, 1914	Eastern Europe, Turkey, Iran	**First World War Eastern Front, Jewish Pale, Armenian Massacres**	Central Total
Feb 4, 1915	South Indian Ocean, Australia, Perth, Darwin, Indonesia, Papua New Guinea	Uninvestigated	Central Annular
Aug 10, 1915	Pacific	Remote	Central Annular
Feb 13, 1916	Pacific, Columbia, Venezuela, Atlantic, off Dublin Coast	**Dublin uprising**	Central Total
Jul 15, 1916	Australia, Indonesia	Uninvestigated	Central Annular
Dec 24, 1916		Partial	Partial 1%
Jan 23, 1917	North Africa, Middle East, Russia	**Fall of Russian Tsar**	Partial 75%
Jun 19, 1917	Canada, Greenland, Russia	**Failure of Mensheviks**	Partial 47%
Jul 19, 1917	South Indian Ocean	Remote, Partial	Partial 9%
Dec 14, 1917	South Atlantic, South Indian Ocean	Remote No Impact	Central Annular
Jun 8, 1918	Pacific, USA	Uninvestigated	Central Total
Dec 3, 1918	Pacific, Chile, Argentina, Atlantic	Uninvestigated	Central Annular
May 29, 1919	Bolivia, Brazil, Atlantic, Gabon, Congo, Zaire, Tanzania	Uninvestigated	Central Total
Nov 22, 1919	Houston, Cuba, Senegal, Mali	Uninvestigated	Central Annular
May 18, 1920	South Indian Ocean, Australia	**Birth of Pope John Paul II**	Partial 97%
Nov 10, 1920	Canada, Ireland	**Irish Independence?**	Partial 74%
Apr 8, 1921	Atlantic, Europe (100% above UK), Russia	**Irish Independence? Russian Civil War?**	Central Annular
Oct 1, 1921	Southern South America, Antarctica	Uninvestigated	Central Total
Mar 28, 1922	Peru to North Africa and Arabia	**Palestinian Mandate**	Central Annular
Sep 21, 1922	Ethiopia, Indian Ocean, Indonesia, Australia	Uninvestigated	Central Total
Mar 17, 1923	Chile, Argentina, Falklands, Atlantic, Namibia, Botswana, Zimbabwe, Mozambique, Madagascar	Uninvestigated	Central Annular
Sep 10, 1923	Off Kamchatka, Pacific, Mexico, Caribbean	Uninvestigated	Central Total
Mar 5, 1924	South Atlantic	Remote	Partial 58%
Jul 24, 1924	South Pacific	Remote, Partial	Partial 17%
Aug 30, 1924	Arctic, North Russia	Remote, Partial	Partial 47%
Jan 24, 1925	Great Lakes, New York, Atlantic, Above UK	Uninvestigated	Central Total
Jul 20, 1925	New Zealand, South Pacific	Uninvestigated	Central Annular

167

Jan 14, 1926	Central African Republic, Uganda, Kenya, Indian Ocean, Indonesia, Philippines	Uninvestigated	Central Annular
Jul 26, 1926	Pacific	Uninvestigated	Central Total
Jan 23, 1927	New Zealand, Pacific, Argentina, Uruguay	Uninvestigated	Central Annular
Jun 29, 1927	Through UK	**UK - From Empire to Welfare**	Central Total
Dec 24, 1927	Antartica, South Pacific, South Atlantic	Remote	Partial 55%
May 19, 1928	South Atlantic	Remote	Non central Total
Jun 17, 1928	North Russia	Remote, Partial	Partial 4%
Nov 12, 1928	Eastern Europe, Russia, Stans	Uninvestigated	Partial 87%
May 9, 1929	South Indian Ocean, Indonesia, Philippines	Uninvestigated	Central Total
Nov 1, 1929	Atlantic, West Africa, Mauritania, Ghana, Central Africa, Zaire, Tanzania	Uninvestigated	Central Annular
Apr 30, 1930	Pacific, California, Canada, North Atlantic	Uninvestigated	Central Annular
Oct 21, 1930	Pacific	Remote	Central Total
Apr 18, 1931	Russia, China	**Japan usurps Manchuria**	Partial 51%
Sep 12, 1931	North Pacific	Remote, Partial	Partial 5%
Oct 11, 1931	South America, South Atlantic, Antarctica	Uninvestigated	Partial 90%
Mar 7, 1932	Antarctica, South Indian Ocean, Australia	Uninvestigated	Central Annular
Aug 31, 1932	Kamchatka, Pacific, Canada	Uninvestigated	Central Total
Feb 24, 1933	Chile, Argentina, Atlantic, Congo, Sudan, Ethiopia, Djibouti, South Yemen	Uninvestigated	Central Annular
Aug 21, 1933	Iran, India	Uninvestigated	Central Annular
Feb 14, 1934	Indonesia, Pacific, off Alaskan coast	Uninvestigated	Central Total
Aug 10, 1934	Atlantic, Angola, Zimbabwe, Mozambique	Uninvestigated	Central Annular
Jan 5, 1935		Partial	Partial 0%
Feb 3, 1935	USA, Mexico, Caribbean	Uninvestigated	Partial 74%
Jun 30, 1935	Russia, Northern Europe, Greenland	Uninvestigated	Partial 34%
Jul 30, 1935	South Atlantic	Remote, Partial	Partial 23%
Dec 25, 1935	South Atlantic, South Pacific	Remote	Central Annular
Jun 19, 1936	Europe, Russia, Asia	**2nd World War Eclipse. Japan invades China**	Central Total
Dec 13, 1936	Australia, South Pacific	Uninvestigated	Central Annular
Jun 8, 1937	Pacific, Peru	Uninvestigated	Central Total
Dec 2, 1937	Pacific	Connects Japan and US via Hawaii	Central Annular
May 29, 1938	South Atlantic, South America, South Africa	Uninvestigated	Central Total
Nov 21, 1938	North Pacific	Remote	Partial 78%

Eclipse Correlation Table 1801-2027

Apr 19, 1939	North America, Northern and Western Europe	**Western Front WW2**	Central Annular
Oct 12, 1939	New Zealand, South Pacific	Uninvestigated	Central Total
Apr 7, 1940	Pacific, Mexico, Southern USA, Atlantic	Uninvestigated	Central Annular
Oct 1, 1940	Colombia, Brazil, Atlantic, Cape Town	Uninvestigated	Central Total
Mar 27, 1941	South Pacific, Peru, Bolivia, Brazil	Uninvestigated	Central Annular
Sep 21, 1941	Stalingrad (now Volgograd) and Guam	**Location of decisive victories in European and Pacific theatres**	Central Total
Mar 16, 1942	South Pacific, Antarctica	Remote	Partial 64%
Aug 12, 1942	South Indian Ocean	Remote	Partial 6%
Sep 10, 1942	North Canada, Western Europe	Uninvestigated	Partial 52%
Feb 4, 1943	Japan - Alaska	**Eclipse over Japan, loser of Pacific War**	Central Total
Aug 1, 1943	Australia, Indonesia	Uninvestigated	Central Annular
Jan 25, 1944	Pacific, Peru, Brazil, Atlantic, Guinea, Mali, Niger	Uninvestigated	Central Total
Jul 20, 1944	Africa, India, Burma, Vietnam, Philippines, Papua New Guinea	**Post WW2 Asian Independence**	Central Annular
Jan 14, 1945	Cape Town, South Indian Ocean	Uninvestigated	Central Annular
Jul 9, 1945	Northwest USA, Canada, Greenland, Russia (Moscow)	**Cold War**	Central Total
Jan 3, 1946	Antarctica, South Indian Ocean	Remote	Partial 55%
May 30, 1946	South Pacific	Remote	Partial 89%
Jun 29, 1946	Arctic	Remote	Partial 18%
Nov 23, 1946	North East USA, Canada	Uninvestigated	Partial 78%
May 20, 1947	Chile, Argentina, Brazil, Atlantic, Ivory Coast, Nigeria, Zaire, Uganda, Kenya	Uninvestigated	Central Total
Nov 12, 1947	Pacific, Peru, Ecuador, Columbia, Brazil	Uninvestigated	Central Annular
May 9, 1948	Indian Ocean, Thailand, Vietnam, Hong Kong, Shanghai, Korea, below Sakhalin	**Vietnam War, Chinese Communist Revolution, Korean War**	Central Annular
May 1, 1948	Zaire, Uganda, Kenya, South Indian Ocean, New Zealand	Uninvestigated	Central Total
Apr 28, 1949	Europe	Uninvestigated	Partial 61%
Oct 21, 1949	Eastern Australia, New Zealand, South Pacific	Uninvestigated	Partial 96%
Mar 18, 1950	South Atlantic	Remote	Non-central Annular
Sep 12, 1950	North East Russia	Uninvestigated	Central Total
Mar 7, 1951	New Zealand, Pacific, Nicaragua	Uninvestigated	Central Annular

169

Sep 1, 1951	Washington, D.C., Atlantic, Mauritania, Ghana, Angola, Zambia, Mozambique, Madagascar	**Washington takes control of Africa**	Central Annular
Feb 25, 1952	Atlantic, Congo, Central African Republic, Sudan, Arabia, Stans, East Russia	**Sudanese self determination**	Central Total
Aug 20, 1952	Pacific, Peru, Bolivia, Uruguay, Argentina, Atlantic,	Uninvestigated	Central Annular
Feb 14, 1953	Eastern Russia, Mongolia, China	Uninvestigated	Partial 76%
Jul 11, 1953	Arctic	Remote, Partial	Partial 20%
Aug 9, 1953	South Pacific	Remote, Partial	Partial 37%
Jan 5, 1954	South Pacific, South Indian Ocean	Remote	Central Annular
Jun 30, 1954	US, Atlantic, Iceland, North Europe, Baltic, Eastern Europe, Caucasus, Iran, Pakistan, India	Uninvestigated	Central Total
Dec 25, 1954	Atlantic, South Africa, Indian Ocean, Indonesia	Uninvestigated	Central Annular
Jun 20, 1955	Sri Lanka, Bangkok, Laos, Vietnam, Philippines	**Vietnam Divided**	Central Total
Dec 14, 1955	Sudan, Ethiopia, Somalia, Thailand, Vietnam, Taiwan	**Suez Crisis, Sudanese Independence and Civil War, Vietnam War**	Central Annular
Jun 8, 1956	New Zealand, Pacific	Remote	Central Total
Dec 2,1956	Eastern Russia, Mongolia, China	Uninvestigated	Partial 87%
Apr 30, 1957	China, East Russia, Canada	Uninvestigated	Non-Central Annular
Oct 23. 1957	South Pacific, South Indian Ocean	Remote	Non Central Total
Apr 15. 1958	Thailand, Vietnam, Laos	**Vietnam again. Taiwan separated from China**	Central Annular
Oct 12, 1958	Pacific, South American Coast	Uninvestigated	Central total
Apr 8, 1959	Australia, Indonesia	Uninvestigated	Central Annular
Oct 2 ,1959	Western and Central Africa	**"Winds of Change" African Independence**	Central Total
Mar 27. 1960	South Indian Ocean, Australia	Uninvestigated	Partial 71%
Sep 20, 1960	East Russia, Canada, USA	Uninvestigated	Partial 61%
Feb 15, 1961	Southern Europe, South France, Florence, Yugoslavia, Romania, Russia, North Russia	**France, Algeria**	Central Total
Aug 11, 1961	South Pacific, Southern Africa	Uninvestigated	Central Annular

Eclipse Correlation Table 1801-2027

Aug 11, 1961	South Pacific, Southern Africa	Uninvestigated	Central Annular
Feb 5, 1962	Indonesia, Pacific	Uninvestigated	Central Total
Jul 31, 1962	Venezuela, Atlantic, Senegal, Mali, Benin, Togo, Nigeria, Cameroon, Congo, Zaire, Tanzania, Madagascar	**Nigerian Civil War**	Central Annular
Jan 25, 1963	Pacific, Southern South America, South Atlantic, South Africa	Uninvestigated	Central Annular
Jul 20, 1963	Pacific, Canada, North East USA	Uninvestigated	Central Total
Jan 14, 1964	South Atlantic	Remote	Partial 56%
Jun 10, 1964	Australia	Uninvestigated	Partial 75%
Jul 9, 1964	Canada, East Russia	Uninvestigated	Partial 32%
Dec 4, 1964	Japan, Pacific	Uninvestigated	Partial 75%
May 30, 1965	New Zealand, Pacific	Uninvestigated	Central Total
Nov 23, 1965	The Stans, Afghanistan, Pakistan, India, Bangladesh, Burma, Thailand, Cambodia, Vietnam, Indonesia, Papua New Guinea	**Indo-Pakistan war, Cambodian and Vietnam Wars. Indonesia's " Year of Living Dangerously"**	Central Annular
May 20, 1966	West Africa, Libya, Istanbul, Stans, China	**Chinese Cultural Revolution**	Central Annular
Nov 12, 1966	Pacific, Peru/Chile, Bolivia, Argentina, Brazil, Uruguay	**Death of Che Guevara**	Central Total
May 9, 1967	North America, Arctic	Remot	Partial 72%
Nov 2, 1967	South Indian Ocean, Southern Africa	Uninvestigated	Non-Central Total
May 28, 1968	Pacific, Atlantic	Remote	Partial 90%
Sep 22, 1968	Stans, Russia, China	Uninvestigated	Central Total
Mar 8, 1969	South Indian Ocean, Indonesia, Pacific	Uninvestigated	Central Annular
Sep 11, 1969	Pacific, Peru, Bolivia	Uninvestigated	Central Annular
Mar 7, 1970	Pacific, Mexico, Florida, Atlantic	Uninvestigated	Central Total
Aug 31, 1970	Pacific, Papua New Guinea	Uninvestigated	Central Annular
Feb 25, 1971	Europe, Russia	Uninvestigated	Partial 79%
Jul 22, 1971	Kamchatka	Remote	Partial 7%
Aug 20, 1971	Antarctica, Australia	Remote	Partial 51%
Jan 16, 1972	South Atlantic, South Indian Ocean	Remote	Central Annular

171

Jul 10, 1972	Canada, Atlantic, UK	Uninvestigated	Central Total
Jan 4, 1973	Pacific, Chile, Argentina, Atlantic, off Congo Coast	Uninvestigated	Central Annular
Jun 30, 1973	Guyana, Atlantic, Mauritania, Sahara, Sudan, Kenya, Indian Ocean	**Sudan State of Emergency**	Central Total
Dec 24, 1973	Central America, Columbia, Brazil, Atlantic Mauritania, Algeria	Uninvestigated	Central Annular
Jun 20, 1974	South Indian Ocean, Australia	Uninvestigated	Central Total
Dec 13, 1974	Canada, USA, North Atlantic	Uninvestigated	Partial 83%
May 11, 1975	Europe, Russia	Uninvestigated	Partial 86%
Nov 3, 1975	Pacific, Argentina, Atlantic	Uninvestigated	Partial 96%
Apr 29, 1976	Atlantic, Mauritania, Algeria, Libya, Turkey, Caspian, Stans, China	**Polisario Insurgency. Death of Mao Tse-tung**	Central Annular
Oct 23, 1976	Tanzania, Indian Ocean, Australia	Uninvestigated	Central Total
Apr 18, 1977	South Atlantic, Namibia, Zambia, Tanzania, Indian Ocean,	**Angolan Civil War**	Central Annular
Oct 12, 1977	Pacific, Columbia, Venezuela	**Jonestown Massacre**	Central Total
Apr 7, 1978	South Atlantic	Remote No Impact	Partial 79%
Oct 2, 1978	North Russia	Remote No Impact	Partial 69%
Feb 26, 1979	Pacific, Canada, Greenland	Uninvestigated	Central Total
Aug 22, 1979	South Pacific, Southern South America	Uninvestigated	Central Annular
Feb 16, 1980	Atlantic, Angola, Zaire, Tanzania. Indian Ocean, Goa, Calcutta, China	Uninvestigated	Central Total
Aug 10, 1980	Pacific, Peru, Bolivia, Paraguay, Brazil	Uninvestigated	Central Annular
Feb 4, 1981	Australia, New Zealand, Pacific, Off South American Coast	Uninvestigated	Central Annular
Jul 31, 1981	Caucasus (Abkhazia, Chechen), Kazakhstan, Eastern Russia, Pacific	Uninvestigated	Central Total
Jan 25, 1982	Antarctica	Remote	Partial 57%
Jun 21, 1982	South Atlantic	Remote	Partial 62%
Jul 20, 1982	Arctic	Remote	Partial 47%
Dec 15, 1982	Europe, Iran, China	Uninvestigated	Partial 74%
Jun 11, 1983	Indonesia	Uninvestigated	Central Total
Dec 4, 1983	Atlantic, Gabon, Zaire, Uganda, Ethiopia, Somalia	**Reignition of Sudanese Civil war**	Central Annular
May 30, 1984	Pacific, Mexico, New Orleans, Washington, Atlantic, Morocco, Algeria	Uninvestigated	Central Annular
Nov 22, 1984	Indonesia, New Zealand, Pacific	Uninvestigated	Central Total
May 19, 1985	Greenland, Kamchatka, Japan	Uninvestigated	Partial 84%
Nov 12, 1985	Pacific, Argentina, Antarctic	Uninvestigated	Central Total

Eclipse Correlation Table 1801-2027

Apr 9, 1986	Antarctica, Australia	Uninvestigated	Partial 82%
Oct 3, 1986	Atlantic, North America	Uninvestigated	Central Annular
Mar 29, 1987	Argentina, Atlantic, Gabon, Cameroon, Zaire, Central African Republic, Sudan, Djibouti,	**Lord's Resistance Army. Rwandan Genocide. South Sudan**	Central Annular
Sep 23, 1987	Kazakhstan, China, Shanghai, Papua New Guinea, Pacific	Uninvestigated	Central Annular
Mar 18, 1988	Sumatra, Borneo, Mindanao, Pacific, Off Canadian Coast	Uninvestigated	Central Total
Sep 11, 1988	Kenya, Indian Ocean, Australia	Uninvestigated	Central Annular
Mar 7, 1989	North America	Uninvestigated	Partial 83%
Aug 31, 1989	South Indian Ocean	Remote	Partial 63%
Jan 26, 1990	South Atlantic, South America	Uninvestigated	Central Annular
Jul 26, 1990	Russia encircled	**Collapse of Soviet Russia**	Central Total
Jan 15, 1991	Australia, New Zealand, Pacific	Uninvestigated	Central Annular
Jul 11, 1991	Pacific, Mexico, Colombia, Brazil	Uninvestigated	Central Total
Jan 4, 1992	Off Indonesia, Pacific, Off California	Uninvestigated	Central Annular
Jun 30, 1992	Buenos Aires, Atlantic, Southern Africa	Uninvestigated	Central Total
Dec 24, 1992	Pacific. Japan	Uninvestigated	Partial 84%
May 21. 1993	Canada, Northern USA, Greenland, Northern Europe	Uninvestigated	Partial 74%
Nov 13, 1993	South Pacific	Remote	Partial 93%
May 19, 1994	Pacific, USA, Morocco	Uninvestigated	Central Annular
Nov 3, 1994	Pacific, Peru, Bolivia, Argentina, Paraguay, Brazil, South Atlantic, South Africa	Uninvestigated	Central Total
Apr 29, 1995	Pacific, Peru, Colombia, Brazil, Atlantic	Uninvestigated	Central Annular
Oct 24, 1995	Tehran, Afghanistan, Pakistan, India, Burma, Thailand, Cambodia, Vietnam, Indonesia, Philippines, Pacific	Uninvestigated	Central Total
Apr 17, 1996	South Pacific	Remote	Partial 88%
Oct 12, 1996	Europe, Atlantic, East Canada	Uninvestigated	Partial 76%
Mar 9, 1997	China, East Russia, Alaska	Uninvestigated	Central Total
Sep 1, 1997	Australia, New Zealand, Pacific	Uninvestigated	Partial 90%
Feb 26, 1998	Pacific, Columbia, Venezuela, Atlantic	Uninvestigated	Central Total
Aug 22, 1998	Indonesia, Singapore, South East Asia, Australia, New Zealand, Philippines, Pacific	Uninvestigated	Central Annular

Feb 16, 1999	Australia, Indonesia, New Zealand	Uninvestigated	Central Annular
Aug 11, 1999	Atlantic, Europe, UK, France, Germany, Turkey, Iraq, Iran, Pakistan, India	**NATO-Militant Islam wars**	Central Total
Feb 5, 2000	Antarctica, South Atlantic	Remote	Partial 58%
Jul 1, 2000	South Pacific, Tierra del Fuego	Remote	Partial 48%
Jul 31, 2000	West Canada, Northern Russia	Remote	Partial 60%
Dec 25, 2000	USA	**Sept 11 Twin Tower attacks**	Partial 72%
Jun 21, 2001	South Atlantic, Angola, Zambia, Zimbabwe, Mozambique, Madagascar, Indian Ocean	Uninvestigated	Central Total
Dec 14, 2001	Pacific, Hawaii, Costa Rica, Caribbean	Uninvestigated	Central Annular
Jun 10, 2002	East Indonesia, Philippines, Pacific, Mexico	Uninvestigated	Central Annular
Dec 4, 2002	Atlantic, Angola, Zambia, Mozambique, Indian Ocean, Australia	Uninvestigated	Central Total
May 31, 2003	Iceland, Russia, Alaska	Uninvestigated	Central Annular
Nov 23, 2003	Australia	Uninvestigated	Central Total
Apr 19, 2004	Southern Africa	Uninvestigated	Partial 74%
Oct 14, 2004	Eastern Russia	Uninvestigated	Partial 93%
Apr 8, 2005	Pacific, Columbia, Venezuela, Mexico (partial)	**Mexican drug wars?**	Central Annular
Oct 3, 2005	Atlantic, Spain (Madrid), Algeria, Tunisia, Libya, Sudan, Kenya, Indian Ocean	**South Sudan, Darfur, bisection**	Central Annular
Mar 29, 2006	Brazil, Atlantic, Nigeria, Niger, Libya, Turkey, Caucasus, Abkhazia, Kazakhstan	**Abkhazia, Borat and Kazakhstan**	Central Total
Sep 22, 2006	Guyana, Brazil, Atlantic Ocean, South Indian Ocean	Uninvestigated	Central Annular
Mar 19, 2007	Stans, India, Mongolia, Russia, Eastern Russia	Uninvestigated	Partial 88%
Sep 11, 2007	Southern Africa, Antarctica	Uninvestigated	Partial 75%
Feb 7, 2008	South Pacific, Antarctica	Remote	Central Annular
Aug 1, 2008	Northwest Canada, Greenland, Europe, Russia, India, China	**Europe Financial Crisis. Rise of BRIC economies**	Central Total
Jan 26, 2009	South Indian Ocean, Indonesia, South East Asia, Australia	Uninvestigated	Central Annular
Jul 22, 2009	India (Bombay, Delhi), China (Chongqing, Shanghai), Japan, Pacific Ocean	Uninvestigated	Central Total

Eclipse Correlation Table 1801-2027

Date	Location	Event	Type
Jan 15, 2010	Central African Republic, Zaire, Uganda, Kenya India Sri Lanka straits, Burma, Chongqing, Shandong	**D. R. Congo Wars, Chongqing Megalopolis/ Poisoning**	Central Annular
Jul 11, 2010	Pacific, Tierra del Fuego	**Falklands UK-Argentina Clash Revival**	Central Total
Jan 4, 2011	North Africa, Middle East, Southern Europe	**Arab Spring, Eurozone Crisis**	Partial 86%
Jun 1, 2011	East Russia, Northern Canada	Uninvestigated	Partial 60%
Jul 1, 2011	Antarctica	Remote, Partial	Partial 10%
Nov 25, 2011	Antarctica, Southern Africa	Remote, Partial	Partial 90%
May 20, 2012	Tonking, East China, East Japan, Pacific Plate. North California, Texas	**China-Japan-USA Islands disputes. Korean peninsular**	Central Annular
Nov 13, 2012	Australia, New Zealand, Pacific, Argentina	**Argentina demonstrations ?**	Central Total
May 10, 2013	Australia, Indonesia, Pacific	**Indonesia**	Central Annular
Nov 3, 2013	Pacific, Gabon, Congo, Zaire, Uganda, Kenya, Ethiopia, Somalia	**D. R. Congo Wars**	Central Annular
Apr 29, 2014	Australia	Uninvestigated	Non-Central Annular
Oct 23, 2014	USA	**Warm-up for USA's difficulties**	Partial 81%
Mar 20, 2015	Europe, Western Russia	**EU Collapse?**	Central Total
Sep 13, 2015	South Africa, Indian Ocean	**South Africa pressurised**	Partial 79%
Mar 9, 2016	Indonesia, South East Asia, North Australia, Papua New Guinea, Pacific	**Indonesia Again**	Central Total
Sep 1, 2016	Gabon, Congo, Zaire, Tanzania	**D. R. Congo Wars**	Central Annular
Feb 26, 2017	Pacific, Chile, Argentina, Atlantic, Angola, Zaire	**D. R. Congo Wars**	Central Annular
Aug 21, 2017	Pacific, USA, Atlantic	**USA First Strike**	Central Total
Feb 15, 2018	Antarctic	Remote	Partial 60%
Jul 13, 2018	Antarctic	Remote, Partial	Partial 34%
Aug 11, 2018	Greenland, Russia	Remote	Partial 74%
Jan 6, 2019	China, Japan, Pacific	**Japan reduced**	Partial 72%
Jul 2, 2019	Pacific, Chile, Argentina, Atlantic	**Argentina Struggles**	Central Total
Dec 26, 2019	Oman, India, Sri Lanka, Indonesia, Philippines, Pacific, Guam	**Indonesia Again**	Central Annular
Jun 21, 2020	Zaire, Sudan, Ethiopia, Yemen, Pakistan, India, Tibet, China, Taiwan	**D. R. Congo Wars, India-Pakistan, China**	Central Annular
Dec 14, 2020	Pacific, Chile, Argentina, Atlantic	?	Central Total
Jun 10, 2021	Northern America, Greenland, Europe, Russia	**Europe Russia again**	Central Annular

The Wheals of God

Dec 4, 2021	South Atlantic, Southern Indian Ocean	Remote	Central Total
Apr 30, 2022	Pacific, Chile, Argentina	?	Partial 64%
Oct 25, 2022	Europe, Western Russia, Middle East	Europe-Russia again	Partial 86%
Apr 20, 2023	Australia, Indonesia	**Indonesia again**	Central Annular
Oct 14, 2023	North Pacific, Western USA, Central America, Columbia, Brazil	**USA Second Strike**	Central Annular
Apr 8, 2024	Mexico, USA	**USA Third Strike**	Central Total
Oct 2, 2024	Pacific, Southern Chile, Argentina	?	Central Annular
Mar 29, 2025	North Atlantic, Europe, Russia	Europe-Russia again	Partial 94%
Sep 21, 2025	New Zealand, South Pacific	?	Partial 86%
Feb 17, 2026	South Indian Ocean	Remote	Central Annular
Aug 12, 2026	Iceland, North Atlantic, Spain	**Europe blockaded?**	Central Total
Feb 6, 2027	Pacific, Chile, Argentina, Atlantic	?	Central Annular
Aug 2, 2027	Mediterranean, North Africa, Red Sea	**Africa disintegrates or cut off?**	Central Total

In summary here are some laws of solar eclipse paths and the historical process. Clearly removing the moon to avert these negative effects is not a realistic option but scientists now have the opportunity to end wars on the earth by the amelioration of the physical and electromagnetic effects of eclipses, albeit appreciating that they are the equivalent of a ship 560 miles long travelling at 200 miles per hour through our ionosphere on a regular basis. Otherwise we will have to learn to live on planet earth with this knowledge and construct our politics and religions accordingly.

In general solar eclipse paths bring about a diminution or significant change of status of those countries they pass over.

Depending on the strength and stability of those countries they can bring about:

i)   civil war/armed rebellion
ii)   foreign invasion, colonisation and decolonisation
iii)   political dissolution or reconstruction
iv)   falls from power of long-standing rulers and dynasties
v)   atrocities, including genocides and scorched earth environments
vi)   the criminalisation of governments
vii)   the denotion of polarised global combatants
viii) the denotion of winners and losers
ix)   the indication of crucial locations in conflicts
ix)   the exacerbation and prolonging of existing conflicts
x)   self-inflicted destruction
xi)   the throwing up of populist deranged messianic leaders
xii)   significant changes of political direction

Additionally eclipse paths create lines of force which attract military conflict to them thereby indicating where fighting is likely to be directed and national boundaries should be redrawn to ensure the peace.

Some eclipses are era-making whilst others are one-offs.

Eclipse effects can be cumulative and can invoke the latent karma from earlier conflicts.

Eclipses are also associated with the births and deaths of historically important individuals as well as the publishing of important intellectual ideas.

Related Websites

www.eclipse.gsfc.nasa.gov is NASA's website, containing maps and tables of 5,000 years of solar eclipses. Whilst its general eclipse graphics are not as clear as those provided by Astrolabe they have some excellent representations of forthcoming eclipses.

www.gcaptain.com/moon-shadow-moon-shadow-moon-moon-shadow-tides/ is one of a number of websites reporting how acoustic waves act as if they are being ploughed by a giant ship during a solar eclipse following a research article that appeared in the journal Geophysical Research in September, 2011.